FLORA ZAMBESIACA

Flora terrarum Zambesii aq̶...

VOLUME TEN: PART ONE

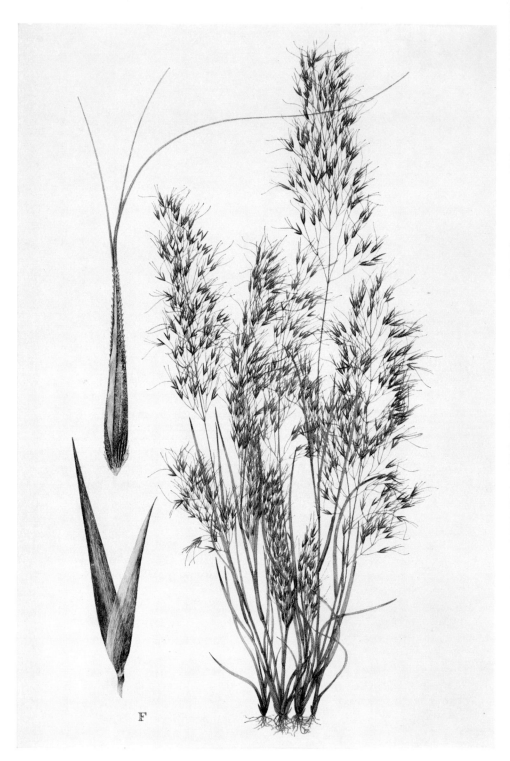

ARISTIDA CUMINGIANA

FLORA ZAMBESIACA

MOZAMBIQUE

MALAWI, ZAMBIA, RHODESIA

BOTSWANA

VOLUME TEN: PART ONE

Edited by

A. FERNANDES, E. LAUNERT & H. WILD

on behalf of the Editorial Board

J. P. M. BRENAN
Royal Botanic Gardens, Kew

A. W. EXELL
Commonwealth Forestry Institute, Oxford

A. FERNANDES
Junta de Investigações do Ultramar, Lisbon

H. WILD
University of Rhodesia

Published by the Managing Committee on behalf of
the contributors to Flora Zambesiaca
Portugal, Malawi, Zambia and
Southern Rhodesia through the
Crown Agents for Oversea Governments and Administrations
4 Millbank, London, S.W.1
December 17, 1971

ISBN 0 85592 014 9

*Printed at the University Press Glasgow
by Robert MacLehose & Company Limited*

CONTENTS

LIST OF FAMILIES INCLUDED IN
VOL. X, PART 1

ANGIOSPERMAE

200. Gramineae (Bambuseae–Pappophoreae)

LIST OF TRIBES REPRESENTED IN THE FLORA ZAMBESIACA AREA

I.	BAMBUSEAE	XV.	ARUNDINEAE
II.	PHAREAE	XVI.	ARISTIDEAE
III.	OLYREAE	XVII.	CENTOTHECEAE
IV.	ORYZEAE	XVIII.	PAPPOPHOREAE
V.	PHYLLORHACHIDEAE	XIX.	DANTHONIEAE
VI.	EHRHARTEAE	XX.	ERAGROSTIDEAE
VII.	POEAE	XXI.	CHLORIDEAE
VIII.	BROMEAE	XXII.	LEPTUREAE
IX.	BRACHYPODIEAE	XXIII.	ZOYSIEAE
X.	TRITICEAE	XXIV.	ARUNDINELLEAE
XI.	MELICEAE	XXV.	ISACHNEAE
XII.	AVENEAE	XXVI.	PANICEAE
XIII.	PHALARIDEAE	XXVII.	ANDROPOGONEAE
XIV.	AGROSTIDEAE	XXVIII.	MAYDEAE

SELECTED BIBLIOGRAPHY OF GENERAL
WORKS RELEVANT TO AFRICAN GRAMINEAE

Taxonomic works of restricted scope will be found cited in their appropriate places in the general text.

F. W. Andrews, *Gramineae* in *The Flowering Plants of the Sudan* **3**: 372–558 (1956).

A. V. Bogdan, *A Revised List of Kenya Grasses* (1955).

N. L. Bor, *Grasses of Burma, Ceylon, India and Pakistan* (1960).

N. L. Bor, *Gramineae*. (Flora of Iraq, **9** (1968)).

N. L. Bor, *Gramineae*. (Flora Iranica, **70** (1970)).

L. Chippindall, A Guide to the Identification of Grasses in South Africa, in D. B. D. Meredith, *Grasses and Pastures of South Africa* (1955).

W. D. Clayton, *A Key to Nigerian Grasses*. (Samaru Research Bulletin No. **1** (1966)).

W. D. Clayton, *Gramineae*, part 1. (Flora of Tropical East Africa (1970)).

D. C. Edwards & A. V. Bogdan, *Important Grassland Plants of Kenya* (1951).

W. J. Eggeling, *An Annotated List of the Grasses of the Uganda Protectorate*, ed. 2. (1947).

F. Gomes Pedro, *Contribuições para o Inventário Florístico de Moçambique, I—Pteridófitas—Monocotiledóneas* in Bol. Soc. Estud. Moçamb. **24**, 87: 1–47 (1954).

K. W. Harker & D. M. Napper, *An Illustrated Guide to the Grasses of Uganda* (1960).

C. E. Hubbard & W. E. Trevithick, *East African Pasture Plants* (1926–1927).

H. Jacques-Félix, *Les Graminées (Poaceae) d'Afrique Tropicale 1–*. (Institut de Recherches Agronomiques Tropicales et des Cultures Vivrières (1962)).

G. Jackson & P. O. Wiehe, *Annotated Check list of Nyasaland Grasses* (1958).

J. Koechlin, *Graminées*. (Flore du Gabon No. **5**, Paris (1962)).

E. Launert, *Gramineae* in Merxmüller, Prodromus einer Flora von Südwestafrika (1971).

F. A. McClure, *The Bamboos*. (Harvard University Press, Cambridge, Massachusetts (1966)).

C. R. Metcalfe, *Anatomy of the Monocotyledons, 1 Gramineae* (1960).

D. M. Napper, *Grasses of Tanganyika* (1965).

R. Pilger, *Gramineae III (Panicoideae)*. (A. Engler & K. Prantl, Die Natürlichen Pflanzenfamilien, ed. 2, **14e** (1940)). *Gramineae II (Micrairoideae, Eragrostoideae, Oryzoideae, Olyroideae)*. (A. Engler & K. Prantl, Die Natürlichen Pflanzenfamilien, ed. 2, **14d** (1956)).

R. Pilger, *Das System der Gramineae*. (A. Engler, Botanische Jahrbücher, **76**: 281–384 (1954)).

A. B. Rendle, *Gramineae*, in *Catalogue of the African Plants Collected by Dr. Welwitsch*, **2,** 1: 135–257 (1899).

W. Robyns, *Flore Agrostologique du Congo Belge et du Ruanda-Urundi, I. Maydées et Andropogonées* (1929); *II. Panicées* (1934). For Latin diagnoses and citation of material of forms described in this work see W. Robyns, Bulletin du Jardin Botanique de l'État, Bruxelles, **8**: 209–243 (1930).

W. Robyns & R. Tournay, *Flore du Parc National Albert*, **3**: 26–210 (1955).

D. P. Stanfield, *Grasses* in Stanfield & Lowe, The Flora of Nigeria (text plus illustrated supplement), Ibadan (1970).

O. Stapf, *Gramineae*. (W. H. Harvey & O. W. Sonder, Flora Capensis, **7** (1898–1900)).

O. Stapf, *Gramineae*. (Flora of Tropical Africa, **9**, 1–4: (1917–1934)).

O. Stapf & C. E. Hubbard, *Gramineae*. (Flora of Tropical Africa, **9**, 4–6 (1920–1934); op. cit. **10**: 1–192 (1937), incomplete).

K. E. Sturgeon, *A Revised List of the Grasses of Southern Rhodesia*. (Rhodesia Agricultural Journal, **50**, 4: 278–291; 5, 420–439; 6, 497–514 (1953); **51**, 1: 12–27; 2, 131–145; 3, 212–226; 4, 293–312; 5, 379–393; 6, 495–508 (1954); **52**, 1, 39–62 (1955); **53**, 1, 110–141 (1956). Also reissued as a reprint (Bulletin Nos. 1724, 1739, 1749, 1755, 1763, 1772, 1781, 1801, 1809, 1872. Ministry of Agriculture, Salisbury, Rhodesia)).

GRAMINEAE

By E. Launert

Inflorescence usually terminal on the culms, rarely axillary, consisting of numerous spikelets arranged in spikes, or in racemes or in a panicle; the spikes or racemes either solitary or more often arranged along a common axis, sometimes digitate, often (*Andropogoneae*) subtended by a spatheole (sheath without lamina) and often branched, thus forming a complicated pseudo-panicle in which frequently a sessile spikelet is paired with a pedicelled one. Spikelets (see pl. tab. 1) usually consisting of 2 (rarely more) glumes and 1 to several lemmas distichously arranged along a rhachilla; each lemma is opposed by a usually thinly membranous palea; enclosed between lemma and palea is the flower (the entire unit is termed a floret); glumes and/or lemmas often awned; florets, and sometimes the entire spikelet, with a basal indurated callus which is either blunt or acute; flower usually ⚥, sometimes unisexual, rather small, consisting of 2–3 minute lodicules (representing the perianth) stamens and ovary; stamens hypogynous, usually 3, sometimes less, rarely 6 or more, with the 2-thecous anthers either opening lengthwise or with a terminal pore; ovary 1-locular; styles 2, rarely 1 or 3; stigmas usually plumose. Fruit 1-seeded, usually a caryopsis, with the pericarp adnate to the seed or sometimes free (rarely fleshy). Mostly annual or perennial herbs, rarely shrubs or trees (*Bambuseae*), often with rhizomes, sometimes stoloniferous. Culms usually cylindrical, very seldom compressed, jointed with the nodes solid, the internodes hollow or sometimes solid, erect, ascending or prostrate, sometimes rooting at the nodes, branched or simple; branches at the base with a hyaline 2-keeled prophyll. Leaves solitary, alternate in 2 ranks, all crowded around the base of the stem or/and spaced along the culm, in most cases consisting of sheath, ligule and lamina (tab. 1). Leaf-sheaths clasping the culm, usually rather tight, ribbed or smooth, with the margins free or connate to a varying degree and often overlapping, often with a pair of auricles on either side of the mouth. Ligule situated transversely at the junction of sheath and lamina, adaxial, membranous or chartaceous, often reduced to a hairy rim, rarely completely absent in some genera. Leaf-laminae usually linear or linear-lanceolate, more rarely ovate or oblong, rather narrow, generally parallel-nerved but in a few instances tessellate, expanded, plicate, or sometimes rolled, thus appearing to be terete, often filiform, rarely amplexicaul or with a sagittate base, occasionally with the very base contracted into a pseudo-petiole, very rarely articulated with the sheath.

An easily recognizable family of c. 10,000 described spp. arranged in more than 600 genera, common all over the globe.

Key to the Tribes

The characters used in this key are not all applicable to Gramineae outside the Flora Zambesiaca area. A key to the genera is to be found at the end of the family. Although morphologically incorrect, in accordance with common usage in agrostological literature the term *pedicelled* is applied to a stalked spikelet.

1.   Arborescent or shrubby plants, flowering erratically. Culms woody, usually very tall (up to 25 m., 5–12·5 cm. in diameter), hollow but often thick-walled. Leaf-lamina finally disarticulating from the sheath, usually constricted at the base into a pseudo-petiole - - - - - - - - - - - - I. BAMBUSEAE
—   Perennial or annual herbs, rarely shrubby, flowering annually. Culms herbaceous, if woody then never more than 2·5 cm. in diameter. Leaf-lamina not disarticulating from the sheath, usually sessile, if with pseudo-petiole then the other characters not applicable - - - - - - - - - - - - - - 2
2.   Spikelets unisexual, different in size and shape; the sexes in different inflorescences, in different parts of the same inflorescence or mixed. Spikelets 1- or 2-flowered 3
—   Spikelets with all the florets ⚥ or with some ♂ or barren and the rest ⚥ (the ♂ or barren floret (s) either below or above the ⚥ one(s)). Spikelets 1–many-flowered 5

3. Spikelets 1-flowered - - - - - - - - - - - 4
—   Spikelets 2-flowered - - - - - - - - - - 34
4. ♀ spikelets sessile or shortly pedicelled; glumes shorter than the floret, acute or
    apiculate; lemmas chartaceous, inflated, utriculate-urceolate, closed but for a small
    circular pore through which the stigmas emerge, covered with hooked hairs. Leaf-
    laminae with the lateral nerves slanting - - - - - II. PHAREAE
—   ♀ spikelets on long clavate pedicels; glumes much longer than the floret, long-
    caudate; lemma coriaceous, whitish, glossy, glabrous, open. Leaf-laminae with the
    lateral nerves running parallel to the midrib - - - - III. OLYREAE
5. Spikelets typically 2-flowered, falling entire at maturity, usually with the superior
    floret ♀ and the inferior ♂ or barren, if barren then often greatly reduced. Spikelets
    frequently dorsally compressed - - - - - - - - - 33
—   Spikelets 1–many-flowered, disintegrating at maturity above the usually persistent
    glumes, if falling entire then not 2-flowered with the superior floret ♀ and the inferior
    ♂ or barren. Spikelets usually laterally compressed to a varying degree - - 6
6. Lemmas either 9-awned or with 4–6 apical hyaline lobes alternating with 5 awns
                                                     XVIII. PAPPOPHOREAE
—   Lemmas 1–3-awned, entire or 2-lobed - - - - - - - 7
7. Leaf-laminae with distinct transverse veinlets (usually best observed on the lower
    surface) - - - - - - - - - - - - - - 8
—   Leaf-laminae without transverse veinlets - - - - - - - 9
8. Ovary glabrous. Spikelets usually more than 4-flowered, if 1–3-flowered then leaves
    with pseudo-petiole. Lemmas 5–7-nerved - - XVII. CENTOTHECEAE
—   Ovary with the apex distinctly pilose. Spikelets 1–4-flowered. Lemma 3–sub
    5-nerved (rarely fully 5-nerved) - - VII. POEAE p.p. (Pseudobromus)
9. Rhachilla internodes densely penicillate near the apex; the hairs 4–10 mm. long,
    obscuring the lemma. Robust reedy perennials with bamboo-like culms up to 8 m.
    tall. Panicles large, plumose - - - - - XV. ARUNDINEAE
—   Rhachilla internodes glabrous or pilose, if pilose then hairs not longer than 3 mm.
    long. Plants much smaller, not reed-like - - - - - - - 10
10. Spikelets falling off entire at maturity, 1-flowered, arranged solitarily or in groups along
    the rigid persistent axis of a raceme or spike-like panicle; glumes either covered with
    conspicuously thickened straight or hooked hairs or drawn out into a long slender
    awn which is several times as long as the body of the glume   XXIII. ZOYSIEAE
—   Spikelets breaking up at maturity leaving the glumes behind, 1–many-flowered, if
    falling entire then arranged in panicles - - - - - - - 11
11. Spikelets arranged in spikes or racemes, these either solitary, digitate or scattered
    along a common axis - - - - - - - - - - - 12
—   Spikelets arranged in open or contracted panicles, or in globose clusters - 19
12. Spikelets sunk into cavities on alternate sides of the rhachis of a solitary spike 13
—   Spikelets not sunk into cavities of the rhachis - - - - - - 14
13. Rhachis of the spike fragile, disarticulating between the spikelets
                                                     XXII. LEPTUREAE
—   Rhachis of the spike persistent or disarticulating irregularly
                                        XXI. CHLORIDEAE p.p. (Oropetium)
14. Spikelets arranged on opposite sides of the rhachis and facing it laterally. Ovary
    with the apex pilose - - - - - - - - - - - 15
—   Spikelets in secund spikes or racemes, if (rarely) on opposite sides then arranged
    edgeways towards the rhachis - - - - - - - - - 16
15. Spikelets sessile or 1-flowered - - - - - - X. TRITICEAE
—   Spikelets shortly pedicelled, many-flowered - - IX. BRACHYPODIEAE
16. Spikelets with 1♀ floret only, with or without ♂ or barren florets beneath or above it
                                                     XXI. CHLORIDEAE
—   Spikelets with 2–many ♀ florets - - - - - - - - 17
17. Spikes or racemes, solitary or scattered along a common axis
                                                     XX. ERAGROSTIDEAE
—   Spikes or racemes digitate - - - - - - - - - - 18
18. Ligule a membrane, glabrous. Spikelets with 2–3 (rarely 4) ♀ florets usually followed
    by 2 to several ♂ or barren ones - XXI. CHLORIDEAE p.p. (Tetrapogon)
—   Ligule a rim of hairs or short ciliate membrane - - XX. ERAGROSTIDEAE
19. Glumes reduced to minute scales or just represented by a tiny collar at the apex of the
    pedicel. Stamens 6. Spikelets strongly laterally compressed   IV. ORYZEAE
—   Glumes, or at least 1 of them fully developed. Stamens usually 3 - - - 20
20. Spikelets strictly 1-flowered, but sometimes with the rhachilla extended beyond the
    floret - - - - - - - - - - - - - - 21
—   Spikelets 2–many-flowered - - - - - - - - - - 23
21. Lemmas subcylindric, indurated at maturity, with a 3-branched awn or rarely
    (Aristida diminuta) with a simple straight 7–10 mm. long apical awn
                                                     XVI. ARISTIDEAE

—   Lemmas neither subcylindric nor indurated, awnless or awned, in the latter case the awn shorter than 5 mm. long or dorsally inserted - - - - 22

22. Lemmas awnless. Glumes usually shorter than the lemmas and similar in texture. Caryopsis with a loose pericarp - - - - XX. ERAGROSTIDEAE

—   Lemmas awned or awnless. Glumes usually longer than the lemmas and firmer in texture. Caryopsis not with a loose pericarp - - XIV. AGROSTIDEAE

23. Spikelets with only 1 ♀ floret - - - - - - - - - 24

—   Spikelets with 2–many ♀ florets - - - - - - 26

24. Spikelets with 2 florets; the inferior ♂ or barren; the superior ♀. Lemma of the ♀ floret usually awned - - - - XXIV. ARUNDINELLEAE

—   Spikelets with 3 florets; the 1st and 2nd reduced to empty lemmas (sometimes 1 of them completely depressed). Lemma of the 3rd (♀) spikelet awnless   25

25. Superior sterile lemma with a basal hinge-like appendage; both sterile lemmas as long as or longer than the ♀ floret - - - VI. EHRHARTEAE

—   Superior sterile lemma without basal appendage; both sterile lemmas much shorter than ♀ floret - - - - - - XIII. PHALARIDEAE

26. Spikelets 2-flowered, awnless, with at least 1 of the lemmas coriaceous
        XXV. ISACHNEAE

—   Spikelets 3–many-flowered - - - - - - 27

27. Glumes usually as long as the spikelet but always longer than the lowermost lemma, if shorter then lemmas with dorsal or geniculate awns. Lemmas 5–many-nerved
        28

—   Glumes shorter than the spikelet and almost always shorter than the lowermost lemma. Lemmas 1–many-nerved, awnless or with a straight apical awn   29

28. Lemmas dorsally awned (sometimes awnless in *Koeleria* but then with the inflorescence conspicuously pubescent or thinly tomentose).   XII. AVENEAE

—   Lemmas awned between 2 apical lobes - - - XIX. DANTHONIEAE

29. Lemmas 1–3-nerved - - - - - - - 30

—   Lemmas 5–many-nerved - - - - - - 31

30. Spikelets awnless - - - - - - XX. ERAGROSTIDEAE

—   Spikelets awned - - - - - XIX. DANTHONIEAE

31. Lemmas awned; the awns 20–43 mm. long, twisted, becoming entangled with the awns from the other florets - - - - XI. MELICEAE

—   Lemmas awnless or with much shorter straight awns - - - 32

32. Ovary crowned by a pilose fleshy appendage above insertion of the styles
        VIII. BROMEAE

—   Ovary pilose or glabrous, without a fleshy appendage - - VII. POEAE

33. Spikelets usually solitary, sometimes paired, more or less similar. Glumes usually membranous; the inferior usually smaller or sometimes absent. Lemma of the inferior floret usually similar to the superior glume in texture; lemma of the superior floret hardened chartaceous, coriaceous or crustaceous, awnless or rarely with a short straight awn - - - - - - XXVI. PANICEAE

—   Spikelets typically paired, with 1 sessile and the other pedicelled, those of each pair similar or more often dissimilar (the pedicelled sometimes much reduced), rarely with the spikelets all similar. Glumes usually as long as the spikelets and enclosing the florets, somewhat rigid and firmer than the hyaline or membranous lemmas. Lemma of the superior floret often awned; the awn usually geniculate
        XXVII. ANDROPOGONEAE

34. Rhachis of the inflorescence leaf-like, with the margins folding over short racemes inserted along its midrib - - - - V. PHYLLORHACHIDEAE

—   Rhachis of the inflorescence triquetrous or cylindic, but not leaf-like
        XXVIII. MAYDEAE

I. BAMBUSEAE Reichenb.

Bambuseae Reichenb., Consp. Reg. Veg.: 54 (1828).

Inflorescence a raceme or panicle, with the spikelets often arranged in large heads or dense clusters. Spikelets all alike, 1–many-flowered. Glumes usually 2 but often more. Lemmas similar to the glumes but exserted from them, 5–many-nerved, commonly coriaceous, usually awnless but sometimes awned from the apex. Palea, if present, 2-keeled or dorsally rounded. Lodicules usually 3. Stamens 3, 6 or numerous; filaments free or connate to a varying degree. Styles 2 or 3. Caryopsis sometimes with a fleshy pericarp; embryo small, basal. Starch grains simple, angular. Chromosomes rather small, basic number 12. A large tribe of woody grasses, usually shrubs or trees, rarely perennial herbs. Culms usually very tall, solid or hollow, erect or sometimes climbing, bearing sheaths with reduced laminas. Leaf-laminae often articulated with the sheaths, usually constricted into a petiole-like base, often with tessellate venation.

Most bamboos flower gregariously in cycles of a few to many years. Many species have never been seen in flower at all. After flowering the plant usually dies.

A more detailed account of the tribe is given by McClure (1966).

The tribe consists of c. 45 described genera. Bamboos occur in forests and woodland throughout tropical and warm-temperate regions of both hemispheres.

Culms solid or thick-walled; leaves not conspicuously tessellate   **2. Oxytenanthera**
Culms always hollow, if thick-walled then leaves conspicuously tessellate:
  Spikelets pedicelled. Stamens 3. Leaf-laminae conspicuously tessellate, with the filiform apex flexuous. Auricles at the shoulder of leaf-sheaths ciliate **1. Arundinaria**
  Spikelets sessile, arranged in dense clusters. Stamens 6. Leaf-laminae indistinctly tessellate, with the apex pungent:
    Stamens with the filaments connate. Inflorescences globose, spiky. Mouth of sheaths of young leaves long-ciliate -   -   -   -   -   **2. Oxytenanthera**
    Stamens with the filaments free. Inflorescences cupuliform, supported by a pseudo-involucre of bracts which are more than ½ as long as the spikelets **3. Oreobambos**

1. ARUNDINARIA Michx.

Arundinaria Michx., Fl. Bor. Amer. **1**: 73 (1803).

Spikelets pedicelled, 1–several-flowered, all ⚥ or often the uppermost one reduced, linear in outline, cylindrical or laterally compressed, usually somewhat loose; rhachilla disarticulating between the florets, with the joints thickened towards the apex. Glumes 2 (more rarely 1), slightly unequal, much shorter than the florets, persistent, few- to many-nerved, membranous. Lemmas large,

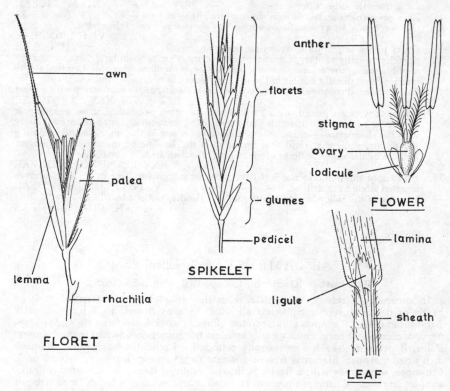

Tab. 1. Schematic diagram showing parts of leaf, spikelet, floret and sexual organs.

membranous, chartaceous or coriaceous, dorsally rounded, 5–13-nerved, usually with transverse veinlets, with the apex acute, acuminate or mucronate. Palea several-nerved, 2-keeled, only little shorter than the lemma, often ciliate along the keels, embraced by the lemma. Lodicules 3, relatively large, ovate-lanceolate, ciliate, with nerves in the lower part. Stamens 3 (very rarely 6). Ovary sub-globular; styles 2 or rarely 3. Caryopsis narrowly oblong in outline, indurate, dorsally rounded, ventrally with a longitudinal furrow; scutellum circular. Shrubby or tree-like bamboos. Culms woody with many prominent nodes. Leaves usually with tessellate venation; laminae finally disarticulating from the sheaths. Inflorescence terminal and/or lateral, variable, paniculate or racemose

A large genus of more than 100 spp. of warm temperate and subtropical regions of Asia, America and (with 4 spp.) Africa. Many species have been cultivated for ornamental purposes.

Arundinaria alpina K. Schum. in Engl., Pflanzenw. Ost-Afr. **C**: 116 (1895).—A. Camus, Bambus.: 48, t.21 fig. b (1913).—Eggeling, Annot. List Grasses Uganda, ed 2: 4 (1947).—Wimbush in E. Afr. Agric. Journ. **13**: 50–56 (1947).—Brenan & Greenway, Check Lists For. Trees & Shrubs **5**, Tanganyika 2: 238 (1949).—Bogdan, List Kenya Grasses: 13 (1951).—Hedberg in Svensk Bot. Tidskr. **45**: 171, 173, 174, 175, 178 (1951).—Robyns & Tournay in Fl. Parc Nat. Alb. **3**: 208, 209 (1955).— Jackson & Wiehe, Annot. Check List Nyasal. Grasses: 30 (1958).—C. E. Hubb. in Hook. Ic. Pl. **36**: t.3594 A & B (1962).—W. D. Clayton in F.T.E.A., Gramineae: 9, fig. 2 (1970). TAB. **2**. Type from Kenya.

A fruticose or arborescent gregarious bamboo with stout creeping woody rhizomes. Culms 2–almost 20 m. tall, 5–12·5 cm. in diam., many-noded, hollow, erect, strict, evenly spaced, green at first later yellow and brown, downy when young later glabrous, profusely branched with several branches from each of the upper nodes; culm-sheaths ovate, oblong or oblong-lanceolate, covered with reddish or reddish-brown stiff hairs, tipped with a linear acute lamina and fimbriate lateral auricles. Leaf-sheaths imbricate, appressed, usually glabrous and smooth, with small deciduous stiffly fimbriate auricles. Ligule c. 2 mm. long, firm, truncate. Leaf-lamina 5–20 × 0·6–1·5 cm., linear-lanceolate to narrowly lanceolate, with the apex acute or somewhat drawn out into a short bristle, base abruptly constricted into a short pseudo-petiole, glaucous to bright green, glabrous, conspicuously tessellate. Panicle (not known from specimens in F.Z. area) terminating the branches and branchlets, 5–15 cm. long, variable, open or condensed; lateral pedicels 1–3 cm. long, the terminal ones much longer. Spikelets 1·5–4 × 0·3–0·4 cm., 5–10-flowered. Glumes (5)7–9-nerved, lanceolate or narrowly ovate, apex acute, chartaceous; the inferior 4–8 mm. long; the superior 5–8 mm. long. Lemmas mostly 7–10 mm. long, 7–9-nerved, acute, finely hairy to glabrous, chartaceous-coriaceous. Palea thinly 7–9-nerved, linear-oblong, with the apex truncate. Lodicules 1·5–2 mm. long. Anthers 3·5–5 mm. long, linear. Caryopsis 4·5–6 mm. long, blackish-brown.

Malawi. C: Dedza Mt., 2140–2230 m., 20.iii.1955, *E. M. & W.* 1083 (BM; LISC; SRGH). **S:** Mt. Mlanje, Luchenya Plateau, 1820 m., 5.vii.1946, *Brass* 16676 (K; SRGH).
Occurs also in Uganda, Kenya, Tanzania, Cameroon, Ethiopia, Sudan, Congo, and in Ruanda. Growing in dense thickets and evergreen forests on mountain slopes, plentiful in the bottom of forested ravines. Usually on deep volcanic humous soil.

Since no flowering material from our area is available the determination of the plant remains doubtful. As already pointed out by C. E. Hubbard (loc. cit.) the specimens of the Malawi bamboo differ from typical *A. alpina* in having thicker-walled culms and in the leaf-laminae lacking the fine bristle-like apex. There is also a remote possibility that our plant may be related to the Drakensberg bamboo, *A. tessellata* Munro. Under this name a specimen has actually been listed by Monro in Proc. Rhod. Sci. Assoc. **6**: 59 (1906). *A. tessellata* itself is by no means closely related to *A. alpina*. Both leaf-shape and the geographic location, however, suggest the attribution of our plant to *A. alpina* rather than to *A. tessellata*.

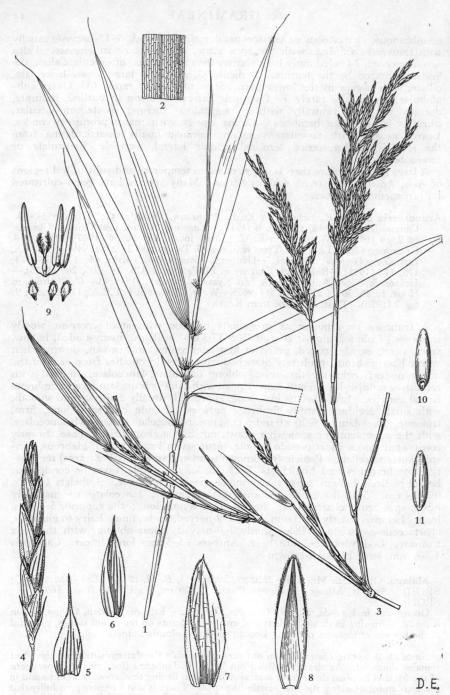

Tab. 2. ARUNDINARIA ALPINA. 1, leafy shoot (× ½); 2, part of underside of leaf-lamina showing tessellate venation (× 2⅔); 3, inflorescence (× ¼); 4, spikelet (× 2⅔); 5, inferior glume (× 5⅓); 6, superior glume (× 5⅓); 7, lemma (× 5⅓); 8, palea (× 5⅓); 9, sexual organs with lodicules separated (× 5⅓); 10, caryopsis, dorsal view (× 5⅓); 11, same, ventral view (× 5⅓), all from *Glover, Gwynne & Samuel* 1187 except 10 to 11 from *Herklots* s.n. From F.T.E.A.

2. OXYTENANTHERA Munro

Oxytenanthera Munro in Trans. Linn. Soc. **26**: 126 (1868).

Spikelets sessile, 1–4-flowered, with the uppermost floret ☿, the lower sterile, terete or slightly laterally compressed, deciduous at maturity; rhachilla-internodes very short, not produced beyond the uppermost floret. Glumes 2 (rarely 3), 17–30-nerved,with transverse veinlets, dorsally rounded, chartaceous to coriaceous; the inferior one shorter than the superior. Lemmas exceeding the glumes, progressively increasing in length towards the apex, 26–32-nerved, with transverse veinlets, dorsally rounded, with the apex pungent. Paleas 16–19-nerved, the ones in the lower florets 2-keeled and dorsally concave, the one in the uppermost floret without keels and convolute. Lodicules usually absent. Stamens 6, with the anthers mucronate; filaments connate. Ovary glabrous; style hollow; stigmas 3. Caryopsis with fused pericarp, narrowly fusiform, crowned with persistent base of the styles. Robust tree-like bamboos, with stout stoloniferous rhizomes. Leaf-laminae finally disarticulating from the sheath. Inflorescence of contracted clusters of spikelets terminating leafy branchlets; each cluster in the axil of a papery sheath with a short lamina.

A monospecific genus confined to the African mainland. Previously 5 Asiatic species were included in *Oxytenanthera* but Holttum, on the ground of its peculiar hollow-style-structure, in his classification (loc. cit.) quite justifiably distinguishes *O. abyssinica* from the others, which are either species of *Dendrocalamus* or *Gigantochloa*.

Oxytenanthera abyssinica (A. Rich.) Munro in Trans. Linn. Soc. **26**: 127 (1868).—
Hack. in Engl. & Prantl, Nat. Pflanzenfam. **2**, 2: 96 (1887).—Engl. in Pflanzenw.
Ost-Afr. **C**: 117 (1895).— Rendle in Cat. Afr. Pl. Welw. **2**, 1: 256 (1899).—Monro in
Proc. Rhod. Sci. Ass. **6**, 1: 60 (1906).—Rendle in Journ. Linn. Soc. **40**: 235 (1911).—
Henkel in S. Afr. Journ. Sci. **24**: 244 (1927).—Eggeling, Annot. List Grass. Uganda,
ed. 2: 30 (1947).—Sturgeon in Rhod. Agric. Journ. **51**: 212 (1954).—Troupin, Fl.
Garamba **1**: 61 (1956).—Holttum in Phytomorphology, **6**: 73 (1965).—Jackson &
Wiehe, Annot. Check List Nyasal. Grass.: 50 (1958).—Napper, Grass. Tangan.:
12 (1965).—W. D. Clayton in F.T.E.A., Gramineae: 11, fig. 3 (1970). TAB. **3**.
Type from Ethiopia.
Bambusa abyssinica A. Rich., Tent. Fl. Abyss. **2**: 439 (1851). Type as above.
Oxytenanthera macrothyrsus K. Schum. in Engl., Pflanzenw. Ost-Afr. **C**: 117
(1895).—Napper, Grasses of Tangan.: 12 (1965). Type from Tanzania.
Oxytenanthera braunii Pilg. in Engl., Bot. Jahrb. **39**: 601 (1907).—Napper, loc.
cit. Type from Tanzania.
Oxytenanthera borzii Mattei in Boll. Ort. Bot. Palermo **8**: 36, 1 (1909). Type from
Eritrea.
Houzeaubambus borzii (Mattei) Mattei in Boll. Soc. Ort. Mut. Soccor. Palermo, **8**,
6: 84 (1910). Type as above.

A robust bamboo, growing in dense clumps. Culms up to 13 m. tall, 5–10 cm. in diam., many-noded, erect, densely covered with appressed hairs when young, later glabrous, solid or thick-walled; culm-sheaths covered with dark-brown stiff hairs; lamina 1–2 cm. long, involute, pungent. Leaf-sheaths without auricles, with a few setae near the collar. Leaf-laminae 5–25 × 1–3 cm., linear-lanceolate to lanceolate, tapering to a fine pungent point, inconspicuously tessellate, glaucous. Spikelet clusters 4–8 mm. in diam., globose, often apparently confluent, spiny; bracts up to 4 mm. long, ovate, dorsally rounded but the lowermost 2 keeled. Spikelets 1·5–4 cm. long, 1–4-flowered, very narrowly lanceolate. Glumes ovate to oblong, apex obtuse to acute, usually scattered with short stiff hairs; the inferior 5–8 mm. long; the superior 8–10 mm. long. Lemmas narrowly lanceolate, dorsally hispidulous; the lower 1·2–2 cm. long; the uppermost almost as long as the spikelet, tapering into a rigid spine up to 7 mm. long. Palea slightly shorter than the lemmas, narrowly lanceolate.

Zambia. N: near Chaporto Village, Isoka, 16.viii.1965, *Lawton* 1251 (FHO). W: Mwinilunga Distr., Matonchi Farm, 18.ii.1938, *Milne-Redhead* 4629 (K; PRE). C: Chakwenga headwaters, 100–129 km. E. of Lusaka, 8.ix.1963, *Robinson* 5644 (K; SRGH). E: Fort Jameson, stream E. of Machinje Hills, 15.v.1965, *Mitchell* 2991 (BM; K; SRGH). S: Mazabuka Distr., Nampeyo area near Siamachenje Village, 19·2 km. down into the escarpment from Chief Chona's Court, headwaters of the Dembela/Lufua rivers, c. 900

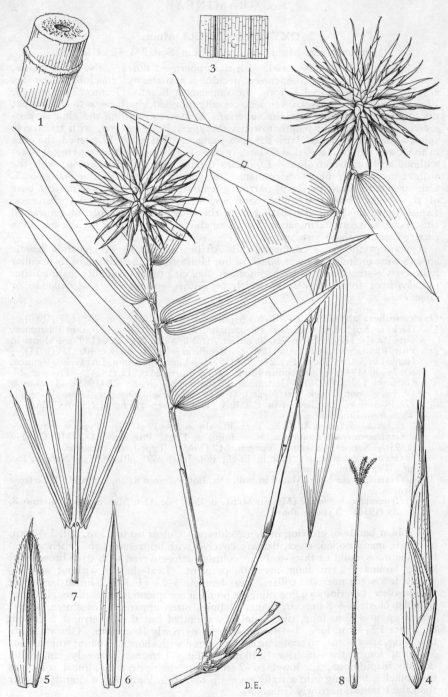

Tab. 3. OXYTENANTHERA ABYSSINICA. 1, part of culm taken from 6 m. above ground-level (× ½); 2, leafy branchlets with inflorescence (× ½); 3, part of underside of leaf-lamina showing tessellate venation (× 5); 4, spikelet (× 2⅓); 5, lemma (× 2⅓); 6, palea (× 2⅓); 7, androecium (× 2⅓); 8, gynoecium (× 2⅓), all from *Balsinhas & Marrime* 343 except 1 from *Snowden* s.n. From F.T.E.A.

m., 15.i.1961, *Bainbridge* 469 (FHO). **Rhodesia.** N: Lomagundi, Sipolilo/Zambezi Valley road, 41·6 km. N. of Sipolilo, v. 1958, *George* 85102 (BM; LISC; PRE; SRGH). W: Victoria Falls, 610 m., 12.ii.1912, *Rogers* 5603 (SRGH). C: Salisbury Experimental Station, 15.iv.1946, *Arnold* 14750 (SRGH). E: Umtali, S. of the road to Beira, 980 m., 9.iv.1961, *Chase* 7455 (BM; SRGH). **Malawi.** C: Lilongwe, *Jackson* 1167 (NYAS). S: S. side of Mt. Mlanje, *Topham* 267 (FHO). **Mozambique.** N: Cabo Delgado, 3 km. from Montepuez towards Natulo, c. 500 m., 7.iv.1964, *Torre & Paiva* 11711 (BR; COI; K; LISC; LM; SRGH). Z: Maganja da Costa, 30 m., 1908, *Sim* 20972 (PRE). MS: Chimoio, Bandula, 8.ix.1956, *Chase* 6191 (SRGH).

Also in Senegal to Ethiopia, Uganda, Kenya, Tanzania, and south to Rhodesia and Mozambique. Growing along banks of perennial watercourses, in damp places at bases of hills, in moorland, on slopes of wooded hills, often on termite mounds.

Note: this species is rather variable with regard to spikelet size, number of fertile florets and thickness of culm walls. J. S. Henkel (loc. cit.) gives a thorough ecological survey of *O. abyssinica* in Rhodesia. According to Eggeling (loc. cit.) flowering takes place over wide areas about every 7th year, the clumps then dying back to sprout again later.

3. OREOBAMBOS K. Schum.

Oreobambos K. Schum. in Notizbl. Bot. Gart. Berl. **1**: 178 (1896).

Spikelets sessile, 2-flowered, terete or somewhat laterally compressed, muticous subtended by a pseudo-involucre of broad coriaceous bracts; florets both ⚥; rhachilla disarticulating beneath the superior glume, produced as a fine bristle beyond the superior floret. Inferior glume absent; superior glume 11–18-nerved, dorsally rounded, chartaceous to coriaceous, with the margins involute. Lemmas exceeding the glumes, many-nerved (11–23), with transverse veinlets, imbricate, dorsally rounded, chartaceous-coriaceous. Paleas slightly shorter than the corresponding lemmas, 5–11-nerved, 2-keeled, flat between the keels. Lodicules absent. Stamens 6; filaments free, anthers distinctly mucronate. Ovary pilose at the apex; stigma 1. Caryopsis crustaceous, with a free pericarp, with a pilose apical appendage. Tall woody bamboos. Leaf-laminae finally disarticulating from the sheath. Inflorescence narrow, slender, large, composed of clusters of spikelets on opposite sides of its branches; each cluster borne in the axil of a deciduous sheath which is usually without a lamina.

A monospecific genus confined to the east African mountains.

Oreobambos buchwaldii K. Schum. in Notizbl. Bot. Gart. Berl. **1**: 178 (1896).—
Eggeling, Annot. List Grass. Uganda, ed. 2: 29 (1947).—Jackson & Wiehe, Annot. Check List Nyasal. Grass.: 49 (1958).—Napper, Grass. Tangan.: 12 (1965).—
W. D. Clayton in F.T.E.A., Gramineae: 13, fig. 4 (1970). TAB. **4**. Type from Tanzania.

A robust tree-like bamboo, growing in small dense patches or solitary clumps. Culms up to 20 m. tall, 5–10 cm. in diam., usually drooping or spreading, rarely erect, green, hollow. Culm-sheaths covered with appressed bristle-like hairs when young, fimbriate at the mouth, with a short subulate lamina. Leaf-sheaths glabrous at the mouth. Leaf-laminae 10–35 × 2·5–6 cm., lanceolate to oblong-lanceolate, very obscurely tessellate, usually glabrous, pale green or somewhat glaucous, apex acute to acuminate. Inflorescence very large, loose; clusters of spikelets 1·5–2 × 0·8–2·25 cm., up to twice their length apart. Bracts 8–14 mm. long, ovate-elliptic, keeled, dark-brown, ciliate along the keel. Spikelets 12–15 mm. long, lanceolate to oblong-lanceolate in lateral view, brown, usually somewhat glossy. Superior glume 9–11 mm. long. Lemmas 10–13·5 mm. long, broadly ovate to elliptic-ovate, with the apex obtuse or subacute. Paleas 8–12 mm. long.

Zambia. N: Sunsu Hill, c. 1950 m., 7.xi.1949, *Rogers in Hoyle* 1359 (FHO; K). **Malawi.** C: Nchisi Mt., c. 1500 m., 4.ix.1929, *Burtt Davy* 1192 (K).

Also known from Uganda and Tanzania. In clearings of evergreen rain forest, in swampy forests and along mountain streams.

Note: not much is known about the flowering habits of this species. Greenway states that it can be seen flowering every year in parts of Tanzania. Nothing has been recorded from elsewhere. None of the specimens studied from the F. Z. area was in flower.

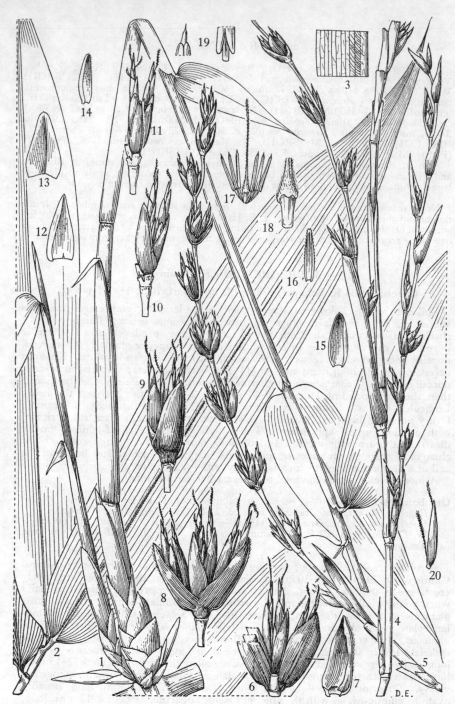

Tab. 4. OREOBAMBOS BUCHWALDII. 1, leafy branchlet (× ⅔) *Hoyle* 1359; 2, leaf-lamina
(× ⅔); 3, part of dorsal surface of leaf-lamina showing tessellate venation (× 4),
both from *Burtt Davy* 1192; 4, 5, part of inflorescence showing spikelet clusters
(× ⅔) *Greenway* 3705; 6, spikelet cluster, ventral view (× 1½); 7, bract removed
from spikelet cluster (× 1½); 8, spikelet cluster dorsal view (× 1½); 9, spikelet
cluster lateral view (× 1½); 10, spikelet cluster partly dissected (× 1½); 11, 2 spikelets
after removal of bracts (× 1½); 12, 13, bracts (× 1½); 14, superior glume (× 1½);
15, lemma (× 1½); 16, palea (× 1½); 17, sexual organ (× 2); 18, detail of ovary (× 6);
19, detail of anther (× 12); 20, terminal floret showing rhachilla extension (× 1½), all
from *Greenway* 2795. From F.T.E.A.

II. PHAREAE Stapf

Phareae Stapf in Harv. & Sond., F.C. **7**: 319 (1898).

Inflorescence a loose stiff few-branched panicle. Spikelets 1-flowered, unisexual, awnless, dissimilar, monoecious. ♂ spikelet smaller than the ♀; glumes 2, small, ovate, 5–7-nerved, usually spreading; lemma cymbiform; palea linear-lanceolate; lodicules 3 or absent; stamens 6, with short filaments and long anthers. ♀ spikelet ovoid to conchiform, more or less inflated; glumes 2, caducous, unequal; lemma chartaceous, much longer than the glumes, either laterally open or closed except for a small pore near the apex, covered with small hooked hairs; palea linear, as long as the lemma, with the apex protruding from the porus; ovary small, gibbous; style 1; stigmas 3, rather long, protruding from the pore of the lemma; caryopsis free within the utriculate lemma; embryo very small; hilum filiform; starch grains simple. Chromosomes small, basic number 12.

A tribe of forest grasses divided into 2 genera. Distributed as for the following genus.

4. LEPTASPIS R. Br.

Leptaspis R. Br., Prodr.: 211 (1810).

Spikelets 1-flowered, monoecious, heteromorphous, unisexual. ♂ spikelets usually much smaller and less conspicuous than the ♀. Glumes 2, much smaller than the spikelet, subequal, membranous, ovate, 5–7-nerved, usually spreading. Lemma similar to the glumes but with more nerves, cymbiform. Palea shorter than the lemma, membranous, distinctly 2-keeled, 2-nerved, narrowly oblong, with the apex bidentate. Stamens 6; filaments rather short; anthers linear. ♀ spikelets subsessile, rather conspicuous, ovoid or conchiform, inflated. Glumes 2, subequal, much shorter than the spikelet. Lemma with the margins connate (thus utricle-like) with an apical or lateral porus, inflated, prominently nerved and usually longitudinally ribbed, more or less densely covered with stiff short hooked hairs. Palea completely enclosed by the lemma, longer than the utricle and often appearing at the porus, very delicate, linear, 2-keeled, with the apex bidentate. Lodicules very small to nil. Ovary small, narrowly ovoid, gibbous; style simple; stigmas 3, rather long, filiform, plumose, exserted through the porus of the lemma. Caryopsis free from the lemma, oblong in outline, compressed; hilum linear, within a longitudinal groove; embryo very small. Perennial rhizomatous bamboo-like grasses. Leaf-laminas large, asymmetric, with a typical slanting venation, tessellate, born on a twisted (180°) pseudopetiole. Ligule scarious, often fringed. Inflorescence a loose panicle, usually terminal, containing male and female spikelets, with the females in a group at the base and the male or males terminal on the branchlets.

A palaeotropical genus of 5–7 spp. represented by one sp. only in tropical Africa and Madagascar.

Leptaspis cochleata Thw., Enum. Pl. Zeyl.: 357 (1864).—Rendle Cat. Afr. Pl. Welw. **2**, 1: 256 (1899); in Journ. Linn. Soc. **40**: 235 (1911).—Sturgeon in Rhod. Agric. Journ. **51**: 212 (1954).—Robyns & Tournay in Fl. Parc Nat. Alb. **3**: 208 (1955).—Pilg. in Engl. & Prantl, Nat. Pflanzenfam., ed. 2, **14d**: 163 (1956).—Bor, Grasses of B.C.I. & P.: 617 (1960).—W. D. Clayton in F.T.E.A., Gramineae: 21, Fig. 7 (1970). TAB. **5**. Type from Ceylon.

Leptaspis conchifera Hack. in Bol. Soc. Brot. **5**: 211 (1887). Syntypes from S. Tomé.

Leptaspis comorensis A. Camus in Bull. Mus. Hist. Nat. Par. **30**: 513 (1924). Type from the Comoro Is.

Stout perennial with a shallow rhizome. Culms 0·3–1·20 m. tall, arising from a decumbent base, producing slender stilt-like roots from the lower nodes. Leaf-sheaths longer than the internodes, tight at first, later somewhat loose and sometimes slipping off the culms, striate, glabrous or puberulous. Ligule short, scarious, ciliolate. Leaf-laminae 10–30 × 2·5–6 cm., oblong to oblong-lanceolate,

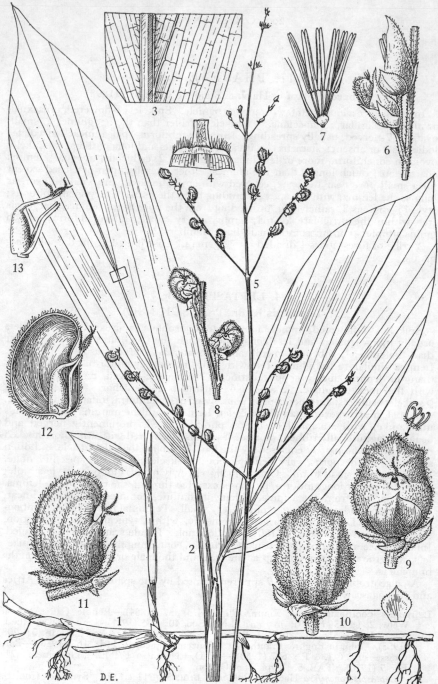

Tab. 5. LEPTASPIS COCHLEATA. 1, base of plant (× ⅔) *Obermeyer* 37159; 2, leafy shoot (× ⅔); 3, part of underside of leaf-lamina showing venation (× 6); 4, detail of false petiole and ligule (× 3); 5, inflorescence (× ⅔); 6, young ♀ and ♂ spikelets (× 6); 7, ♂ sexual organs (× 8); 8, ♀ spikelets (× 2); 9, ♀ spikelet, ventral view showing hooked hairs (× 6); 10, same, dorsal view showing glume (× 6); 11, same, lateral view (× 6); 12, ♀ floret, longitudinal section (× 6); 13, palea and gynoecium (× 6). 2–5 from *Gossweiler* 4666, 6–13 from *Leach & Chase* 10513.

abruptly acute at the apex, asymmetrically attenuate into the 1–2·5(4) cm. long pseudo-petiole, thinly but stiffly chartaceous, green and somewhat glossy above, almost concolorous beneath, with the midrib prominent and whitish beneath, glabrous or scattered pilose, smooth; transverse venation distinct beneath. Panicle up to 45 × 20 cm., loose, with the branches 2–3 in a whorl; rhachis purplish-green, terete, smooth, usually glabrous; branches patent, deflexed or ascending, more rarely appressed to the rhachis. ♂ spikelets c. 4 mm. long; glumes about ½ as long, purplish. ♀ spikelets 4–6·5 mm. long, becoming detached and adhering like burrs by means of hooked hairs; glumes 2–3 mm. long, ovate, acuminate, often purplish. Lemma 4–6 mm. long, conchiform, with prominent ribs; pore lateral. Caryopsis not seen.

Rhodesia. E: Melsetter Distr., Mt. Selinda Forest, 10.i.1949, *Chase* 3923 (BM; SRGH). **Mozambique.** Z: Mts. of Morrumbala, 30.iv.1943, *Torre* 5234 (LISC). MS: Chimoio, Bandula Forest, 910 m., 3.iv.1952, *Chase* 4437 (BM; K; LISC; SRGH).

Throughout tropical Africa, Madagascar, Ceylon, extending to Malaysia and the Solomon Is. In the ground layer of rain-forests or secondary forests; usually in dense patches, locally abundant.

III. OLYREAE Spenner

Olyreae Spenner, Fl. Frib. **1**: 172 (1825).

Plants monoecious. Inflorescence a panicle or raceme. Spikelets unisexual, 1-flowered, heteromorphous, with the 2 sexes either mixed or in separate inflorescences on the same plant; ♀ spikelets much larger than the ♂ and usually positioned above them. Glumes usually similar, often acuminate or caudate, deciduous; minute or suppressed in ♂ spikelets. Lemmas coriaceous or cartilaginous (♀ spikelets) or membranous (♂ spikelets), usually awnless, shorter than the glumes. Lodicules 3 rarely 2. Stamens 3 rarely 2. Caryopsis with the embryo rather small, basal; hilum linear; starch grains compound. Basic chromosome number not yet known.

A tribe of broad-leaved forest grasses comprising 12 genera, mainly distributed throughout tropical America, 1 genus endemic to New Guinea, 1 extending to tropical Africa and the Mascarene Is.

5. OLYRA L.

Olyra L., Syst. Nat., ed. 10, **2**: 1261 (1759).

Spikelets monoecious, 1-flowered, deciduous, heteromorphous, with both sexes usually within the same (very rarely in different) inflorescences of each individual plant; rhachilla disarticulating below the lemma. ♂ spikelets more numerous and much smaller than the ♀, and arranged below them, deciduous. Glumes reduced to a very small, almost invisible rim below the articulation of the rhachilla. Lemma 3-nerved, membranous, broadly lanceolate, with inflexed margins, apex acuminate or shortly awned. Palea slightly shorter than the lemma, 2-nerved, lanceolate, thinly membranous, the margins embraced by the lemma. Lodicules usually 3, cuneate, very small. Stamens 3. Ovary completely absent. ♀ spikelets large, conspicuous, arranged in the upper part of the inflorescence above the ♂ ones, ventricose. Glumes 2, subequal, (3)5–9-nerved, persistent, chartaceous or membranous, transversely veined, ovate to lanceolate, cymbiform, with the apex usually caudate. Lemma faintly 5-nerved, awnless, subcoriaceous to indurate (bony), elliptic to lanceolate, dorsally convex. Palea 2-nerved, similar in shape and consistency to the lemma, tightly embraced by the margins of the lemma. Lodicules 3, cuneate with the apex truncate. Stamens entirely absent. Ovary glabrous; styles fused up to the middle; stigmas plumose, apically exserted. Caryopsis tightly enclosed by the lemma and palea, convex on both sides; hilum almost as long as the caryopsis, linear. Perennial bamboo-like grasses. Leaf-laminae borne on a short pseudo-petiole, relatively large, somewhat asymmetric, convolute in bud, tessellate. Ligules thinly membranous. Inflorescence a terminal panicle (rarely with some additional axillary ones).

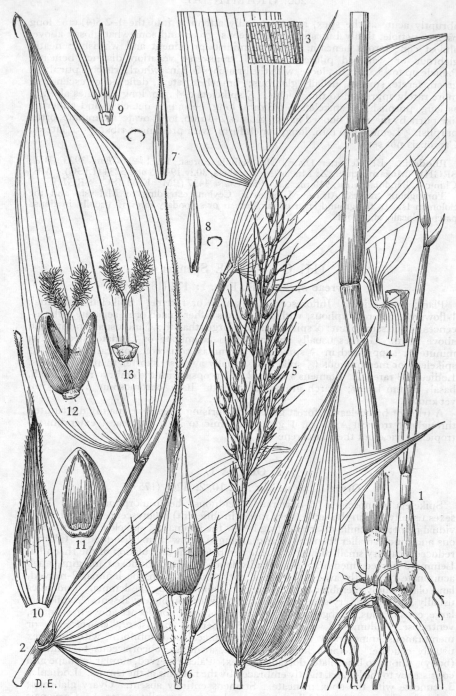

Tab. 6. OLYRA LATIFOLIA. 1, base of plant (×⅔) *Gomes e Sousa* 4375; 2, leafy shoot (×⅔); 3, part of underside of leaf-lamina showing tessellate venation (×2); 4, detail of false petiole and ligule (×2) 2–4 from *Schweinfurth* 7579; 5, inflorescence (×⅔) *Gossweiler* 625; 6, cluster of one ♀ and two ♂ spikelets (×4); 7, lemma of ♂ spikelet (×4); 8, palea of same (×4); 9, ♂ sexual apparatus (×4); 10, inferior glume of ♀ spikelet (×4); 11, ♀ floret (×4); 12, ♀ floret, opened (×4); 13, ♀ sexual organs (×4) 6–13 from *Torre & Paiva* 10460.

A predominantly tropical genus of c. 25 spp. mainly in tropical and subtropical America, with only 1 sp. occurring in Africa and Madagascar.

Olyra latifolia L., Syst. Nat., ed. 10, **2**: 1261 (1759).—Engl., Pflanzenw. Ost-Afr. **C**: 106 (1895).—Rendle in Cat. Afr. Pl. Welw. **2**, 1: 255 (1899).—Stapf in Harv. & Sond., F. C. **7**: 746 (1900).—Wood, Natal Pl. **5**, 3: t. 464 (1906).—R. E. Fr. in Wiss. Ergebn. Schwed. Rhod.-Kong. Exped. **1**: 205 (1916).—Sturgeon in Rhod. Agric. Journ. **51**: 213 (1954).—Eggeling, Annot. List Grass. Uganda: 29 (1947).—Robyns & Tournay in Fl. Parc Nat. Alb. **3**: 206 (1955).—Chippindall in Meredith, Grasses & Pastures of S. Afr.: 453 (1955).—Troupin, Fl. Garamba, **1**: 59 (1956).—Pilg. in Engl. & Prantl, Nat. Pflanzenfam. ed. 2, **14d**: 159 (1956).—Jackson & Wiehe, Annot. Check List Nyasal. Grass.: 49 (1958).—Bogdan, Rev. List Kenyan Grass.: 35 (1958).—Harker & Napper, Ill. Guide Grass. Uganda: 42 (1960).—Napper, Grass. Tangan.: 50 (1965).—W. D. Clayton in F.T.E.A., Gramineae: 17, fig. 6 (1970). TAB. **6**. Type from Jamaica.

Olyra brevifolia Schumach. in Dansk. Vid. Selsk. **4**: 176 (1829). Type from Guinea.

A stout perennial grass. Culms from 0·9–almost 5 m. tall, many-noded, usually branched, erect, straggling or arching, glabrous, usually glossy. Leaf-sheaths striate, rather tight, inconspicuously keeled in the upper part, glabrous or scattered pilose, the mouth always shortly ciliate. Ligule rather short, truncate, often lacerate, broader than the base of the pseudo-petiole. Leaf-laminae 7–20 × 2·5–7 cm., glaucous, ovate, ovate-oblong to lanceolate-oblong, asymmetric to a varying degree, base rounded or broadly cuneate, contracted into the short hairy pseudo-petiole, apex acuminate, expanded, glabrous, or scattered hairy on the lower surface. Panicle (3)7–18 cm. long, terminal (rarely with a laterally arising one from the axil of the uppermost leaf), erect, usually contracted, narrowly triangular in outline; rhachis angular, sometimes hairy; branches simple or divided. Pedicels of the ♂ spikelets filiform, of the ♀ spikelets conspicuously clavate. ♂ spikelets laterally arranged, 3–6 mm. long, linear-lanceolate, acuminate; lemma awned; anthers c. 2·5 mm. long. ♀ spikelets 7–10 mm. long, rather turgid; inferior glume c. 7–9 mm. long, acuminate, 5–7-nerved; the superior almost twice as long, tapering into a long flexuous cauda, 7–9-nerved; floret bony, whitish, glossy, shed without the persistent glumes.

Zambia. N: Luwingu, 7.v.1958, *Fanshawe* 4413 (SRGH). **Rhodesia.** E: Inyanga, Pungwe R., Holdenby, 910 m., 16.iii.1965, *Chase* 8275 (BM; K; SRGH). **Malawi.** S: Port Herald, 23.iii.1960, *Phipps* 2630 (K; SRGH). **Mozambique.** N: Malema, Serra Merripa, 5.ii.1964, *Torre & Paiva* 10460 (BR; COI; K; LISC; LM; SRGH). Z: Milange, Serra Tumbine, near provaoção Vila Masseti, c. 300 m., 18.i.1966, *Correia* 446 (LISC). MS: Chimoio, streamside near Garuso, 610 m., 1.iv.1952, *Chase* 4436 (BM; COI; K; LISC; SRGH). SS: Gaza, Vila de João Belo, Chipenhe, by forest of Chirindzene, 9.vi.1960, *Lemos & Balsinhas* 49 (BM; COI; K; LISC; SRGH).

Also in tropical America, throughout tropical Africa and NE. Cape Prov., Zululand and Madagascar. In mountain forests, riverine forests, along streams and roads in shade.

IV. ORYZEAE Dumort.

Oryzeae Dumort., Obs. Gram. Belg.: 83, 135 (1823).

Inflorescence a panicle, open or sometimes contracted. Spikelets all alike, strongly laterally compressed, ♀, usually 1-flowered but sometimes 3-flowered with the first and second floret reduced to narrow empty lemmas; rhachilla disarticulating beneath the fertile floret. Glumes suppressed or represented by 2 microscopical rims at the apex of the pedicel. Lemma 3–9-nerved, chartaceous to coriaceous, awnless or awned from the apex, sometimes with the apex acuminate to caudate; inferior lemmas, if present, usually scale-like. Palea similar in both texture and size to the lemma. Lodicules 2, entire or 2-lobate. Stamens 6, 3 or 2. Ovary glabrous; styles 2; stigmas plumose, laterally exserted. Caryopsis with the embryo rather small; hilum linear; starch grains compound. Chromosomes small, basic number 12.

A tribe of 7 genera, distributed throughout the tropical and warm temperate regions of the world.

Spikelets with 2 greatly reduced sterile lemmas beneath the fertile floret   **7. Oryza**
Spikelets without sterile lemmas beneath the floret -   -   -   -   **6. Leersia**

6. LEERSIA Sw.

Leersia Sw., Prodr. Veg. Ind. Occ.: 21 (1788) *nom. conserv.*

Spikelets shortly pedicelled or subsessile, 1-flowered, ⚥, laterally compressed, deciduous; rhachilla articulating above the rudimentary glumes.* Sterile lemmas not present; fertile lemma cymbiform, strongly keeled, 5-nerved (one pair of nerves very close to the margins), chartaceous to coriaceous, awnless or tapering in a cauda or pseudo-awn; keels and margins usually stiffly ciliate more rarely scabrous or smooth. Palea narrower and slightly shorter than the lemma but of both the same consistency and indument but with the margins usually hyaline, tightly clasped by the margins of the lemma, apex obtuse, acute, acuminate or drawn out into a short cauda. Lodicules 2, very small, finely nerved. Stamens 6 (in all species of the Flora Zambesiaca area), 3 or rarely 2 with the filaments free. Ovary glabrous; stigmas plumose, laterally exserted. Caryopsis remaining tightly enclosed between lemma and palea, oblong, elliptic-oblong to ovate in outline, laterally compressed, whitish to dark brown in colour, embryo very small; hilum long, linear. Annual or perennial grasses, growing in a variety of moist habitats, often stoloniferous. Culms erect or straggling; nodes almost always hairy to a varying degree. Ligule membranous, entire or sometimes lacerate, usually glabrous. Inflorescence an open or contracted panicle with slender branches; the spikelets often densely imbricate.

A genus of c. 15 spp., distributed throughout the tropics and subtropics of the world; 1 sp. extending into the temperate regions.

Lemma caudate or tapering into a long pseudo-awn:
  Lemma tapering into a slender 110–180 mm. long pseudo-awn   5. *nematostachya*
  Lemma caudate; the cauda (0·8)1·5–7·5(20) mm. long - - - 4. *tisserantii*
Lemma without pseudo-awn and not caudate, usually with a short apical prolongation not
  exceeding 0·8 mm. in length:
  Keels of lemma and palea (and usually the flanks also) with semi-circularly curved
    cilia - - - - - - - - - - - 6. *oncothrix*
  Keels of lemma and palea ciliate, ciliolate (the cilia straight or slightly curved) or
    scabrous; flanks stiffly hispid or glabrous:
    Keels of lemma and palea distinctly ciliate; cilia 0·2–0·6 mm. long or even slightly
      longer, usually stiff - - - - - - - - 1. *hexandra*
    Keels of lemma and palea scabrous or scaberulous or ciliolate and then with the
      ciliola flexible, not exceeding 0·1 mm. in length:
      Nodes silvery villous or densely sericeous. Spikelets (1·2)1·3–1·5 mm. wide; keels
        of lemma and palea as well as margins of lemma shortly ciliolate. Panicle (1)1·5–
        5 cm. wide, ovate-oblong or narrowly to broadly elliptic in outline; the branches
        obliquely ascending or sometimes spreading - - - - 2. *denudata*
      Nodes almost always glabrous. Spikelets 0·9–1(–1·25) mm. wide; keels of the
        lemma and palea as well as margins of lemma scabrous, scaberulous, or very
        rarely ciliolate. Panicle 0·5–1 (very rarely –1·5) cm. wide, linear, linear-oblong
        or rarely narrowly elliptic in outline; the branches usually appressed to the
        rhachis - - - - - - - - - - - 3. *friesii*

1. **Leersia hexandra** Sw., Prodr. Veg. Ind. Occ.: 21 (1788).—Stapf in Harv. & Sond., F.C. **7**: 659 (1900).—R. E. Fries, Wiss. Ergebn. Schwed. Rhod.-Kong. Exped. **1**: 205 (1916).—Stent in Bothalia, **1**, 4: 275 (1925).— C. E. Hubbard & Trev., East Afr. Pasture Pl. **1**: 46 (1926).—Eggeling, Annot. List Grass. Uganda ed. **2**: 26 (1947).— Bogdan & Edwards, Import. Grassl. Pl. Kenya: 47 (1951).— Sturgeon, Rev. List Grass. S. Rhod. in Rhod. Agric. Journ. **51**: 213, 217 (1954).—Robyns & Tournay, Fl. Parc Nat. Alb. **3**: 205 (1955).—Chippindall in Meredith, Grasses & Pastures of S. A.: 33 (1955).—Pilg. in Engl. & Prantl, Nat. Pflanzenfam. ed. 2, **14d**: 151, 152 (1956).—F. W. Andr., Fl. Pl. Sudan, **3**: 479 (1956).—Troupin, Fl. Garamba, **1**: 51 (1956).—Jackson & Wiehe, Annot. Check List Nyasal. Grass.: 46 (1958).—Bogdan, Rev. List Kenya Grass.: 35 (1958).—Bor, Grasses of B.C.I. & P.: 599 (1960).— Harker & Napper, Illustr. Guide Grass. Uganda: 39 (1960).—Napper Grass. Tangan.: 50 (1965).—Launert in Senckenb. Biol. 46: 140, fig. 4, 19–20 (1965).— Berhaut, Fl. Sénégal: 403 (1967).—W. D. Clayton in F.T.E.A., Gramineae: 25, fig. 9 no. 11, (1970).—Launert in Merxm., Prodr. Fl. SW. Afr. 160: 125 (1970). TAB. 7 fig. C. Type from Jamaica.
    Asprella hexandra (Sw.) Roem. & Schult. in L., Syst. Veg. **2**: 267 (1817). Type as above.

* See footnote p. 31.

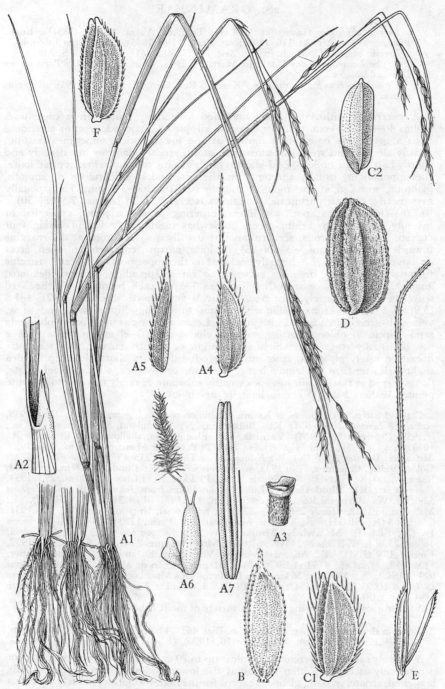

Tab. 7. A.—LEERSIA TISSERANTII. A1, habit (×⅔); A2, ligule (×6); A3, pedicel (×12); A4, lemma (×6); A5, palea (×6); A6, ovary with stigma and one lodicule (×15); A7, anther (×18) all from *McCallum-Webster* A155. B.—LEERSIA FRIESII, spikelet (×9) *Siame* 624. C.—LEERSIA HEXANDRA. C1, spikelet (×9); C2, caryopsis (×12) both from *Jackson* 24. D.—LEERSIA ONCOTHRIX, spikelet (×9) *Vesey-FitzGerald* 3748. E.—LEERSIA NEMATOSTACHYA, spikelet (×4) *Robinson* 5897. F.—LEERSIA DENUDATA, spikelet (×9) *Robinson* 4345.

Leersia abyssinica Hochst. ex A. Rich., Tent. Fl. Abyss. **2**: 356 (1851).—Engl.,
Pflanzenw. Ost-Afr. **A**: 116 (1895); op. cit. **C**: 106 (1895). Type from Ethiopia.
Leersia capensis C. Muell. in Bot. Zeit. **14**: 345 (1856). Type from S. Africa.
Oryza hexandra (Sw.) Doell. in Mart. Fl. Bras. **2**, 2: 10 (1871). Type as for
Leersia hexandra.
Homalocenchrus hexandrus (Sw.) Kuntze, Rev. Gen. Pl. **2**: 777 (1891). Type as
above.

A perennial with long stout branched rhizomes, more rarely caespitose.
Culms 40–85(–125) cm. tall, many-noded, simple or branched, erect or ascending
from a geniculate base, often rooting at the lower nodes, sometimes floating,
usually slender and weak but sometimes stout, striate, glabrous but densely and
shortly hairy at the nodes. Leaf-sheaths finely striate, terete, the lowest often loose,
the upper tight, usually shorter than the internodes, scaberulous or smooth,
glabrous, auricled at the mouth. Ligule (0·9)1·5–2·5(–3·4) mm. long, usually
asymmetric, obliquely truncate, sometimes lacerate. Leaf-laminae 7·5–18(–30) ×
(0·2)0·3–0·8(–1·6) cm., narrowly linear, tapering to a fine point, expanded or
involute, glaucous to bright green, somewhat rigid, rarely flaccid, straight or
curved, usually glabrous, scaberulous but usually very scabrous at the margins
towards the base. Panicle 5–10(–15) cm. long, narrow, contracted to open, erect
or somewhat nodding, often partly enclosed by the uppermost leaf-sheath; rhachis
furrowed, scabrous; branches suberect or rarely spreading, filiform, flexuous,
angular, scaberulous or smooth. Racemes 1–5 on each branch, with the 3–10
spikelets usually densely imbricate. Pedicels very short. Spikelets (3·2)3·4–4·8
(5·9) × 1·2–1·4(–1·7) mm., obliquely oblong to oblong-elliptic in lateral view,
yellowish-green, often tinged with purple. Lemma as long as the spikelet, obliquely
semi-elliptic to oblong, cartilaginous to chartaceous, keel and margins with a
row of usually stout cilia which are straight or slightly curved, flanks slightly
hispid or rarely glabrous, apex usually produced into a short blunt tip. Palea
slightly shorter than the lemma but of the same consistency, with the keel ciliate,
flanks hispid or rarely glabrous, apex usually subacute or rarely tapering to an acute
point. Anthers 2–2·8(–3·2) mm. long, linear-oblong.

Caprivi Strip. 11·2 km. S. of Katima Mulilo on road to Ngoma, 910 m., 22.xii.1958,
Killick & Leistner 3030 (BM; K). **Botswana.** N: Ngamiland, Matlapaneng, 940 m.,
iv.1959, *Robertson* 669 (BM). **Zambia.** B: Shangombo, shallow water in Mashi R.,
1040 m., 8.viii.1952, *Codd* 7452 (K). N: Mningi Pans near Abercorn, 1710 m., 21.ii.1959,
McCallum Webster A116 (K). W: Ndola, Itawa Dambo, 25.v.1950, *Jackson* 23 (K). C:
Golden Valley, Chisamba, 17.iii.1933, *Michelmore* 694 (K). E: Lundazi, 1000 m., 1.vi.1954,
Robinson 800 (K; SRGH). S: Muckle Neuk, 19·2 km. N. of Choma, 1280 m., 27.ii.1954,
Robinson 562 (K). **Rhodesia.** N: Lutopi vlei c. 48 km. from Gokwe, 18.iii.1963, *Bingham*
529 (K). W: Shangani Res., Cockcroft's Pan, 1.iv.1951, *West* 3164 (K; SRGH). C:
Makwiro, iii.1922, *Mundy* 2867 (K; SRGH). E: Cashel, Umvumvumvu R., 16.ii.1951,
Crook 368 (K; SRGH). S: Victoria, Makoholi Expt. Farm, 1220 m., 14.i.1948, *Robinson*
195 (K; SRGH). **Malawi.** N: Misuku, 28.vi.1951, *Jackson* 564 (K). S: Lake Shilwa,
460 m., 23.vi.1949, *Wiehe* N/145 (K; SRGH). **Mozambique.** Z: Gúruè, 610 m., 10.ii.1932,
Vincent 179 (BM). T: Angonia between Vila Coutinho and the Angonia frontier,
11.v.1948, *Mendonça* 4153 (LISC). MS: Manica, região de Mavita, 30.i.1948, *Barbosa*
924 (LISC). N: Muecate, 36 km. from Imala towards Mocuburi, 450 m., 16.i.1964, *Torre
& Paiva* 10025 (COI; LISC; LM; PRE). LM: near Namaacha, 20.xii.1944, *Torre*
6894 (K; LISC).

A pantropical grass growing in a large variety of moist habitats.

2. **Leersia denudata** Launert in Senckenb. Biol. **46**: 144, fig. 23 (1965).—W. D. Clayton,
in F.T.E.A., Gramineae: 27, fig. 9 no. 10, (1970). TAB. **7** fig. F. Type from Tanzania.

A densely caespitose perennial. Culms up to 70 cm. tall, 4–many-noded, erect or
geniculately ascending, often rooting at the lower nodes, slender, terete, usually
branched, striate, smooth, glabrous except for the nodes which are densely sericeous
or villous. Leaf-sheaths striate, tight when young, later loose and slipping off the
culm, slightly shorter than the internodes, auricled at the mouth, smooth, glabrous,
or scattered with very short hairs. Ligule c. 1·5 mm. long, obliquely truncate,
slightly retuse. Leaf-laminae 3–9(–14) × 0·15–0·4(–0·6) cm., narrowly linear,
acute or tapering to a fine point, usually expanded or folded about the midrib,
rarely involute, pale green, flaccid, densely arranged along the culm, glabrous,
smooth on the upper surface. Inflorescence as described for *L. hexandra,* usually

salmon-pink or purple in colour. Spikelets (3·2)3·6–4·4(–5·5) × (1·2)1·3–1·5 mm., similar in shape to *L. hexandra*. Lemma usually with an acute or subacute apical prolongation slightly longer than observed in *L. hexandra*, with the keels and margins shortly ciliolate (often only towards the apex) or more rarely scabrous. Palea acute or with a prolongation akin to that of the lemma, with the keel ciliolate (often only in the upper part). Sexual apparatus and caryopsis as in *L. hexandra*.

Zambia. B: Bulozi, 31.xii.1959, *Gilges* 801 (SRGH). N: Kasama Distr., 16 km. E. of Kasama, 4.ii.1961, *Robinson* 4354 (K; SRGH). W: Mufulira, 11.v.1934, *Eyles* 8382 (K; SRGH). C: Chakwenga headwaters, 100–129 km. E. of Lusaka, 16.xi.1963, *Robinson* 5816 (SRGH). S: Kafue Flats, Mazabuka, 13.iv.1962, *Mitchell* 13/100 (K; SRGH). **Rhodesia.** N: Umboe Flats, 1100 m., 12.iii.1931, *Brain* 2790 (SRGH). C: Salisbury, on road to Norton, 11.ii.1929, *Young* 5 (K). E: Imbeza Estate, iii.1935, *Gilliland* 1741 (BM; SRGH). S: Buhera, Mwerihari R., 910 m., v.1953, *Davies* 360 (SRGH).

Also in Kenya, Tanzania, Transvaal and perhaps Lesotho. Forming usually dense masses on fringes of ponds and lakes, or large glades in swamp grassland on black alluvial soil. Often appearing in pure stands covering 100 acres or more.

3. **Leersia friesii** Melderis in Svensk Bot. Tidskr. **40**: 225 (1946).—Harker & Napper, Ill. Guide Grass. Uganda: 39 (1960).—Napper, Grass. Tangan.: 50 (1965).— Launert in Senckenb. Biol. **46**: 146, fig. 6, 18 (1965).—W. D. Clayton, F.T.E.A. Gramineae: 27, fig. 9 no. 9 (1970). TAB. 7, fig. B. Type from the Congo Republic.

A rhizomatous perennial. Culms up to 80 cm. tall, 3–6(8)-noded, erect or more often geniculately ascending, branched from the lower nodes or rarely simple, spongy, striate, smooth, glabrous, slightly bulbous at the lower-most nodes; nodes glabrous. Leaf-sheaths rather loose; the lower ones scale-like, blade-less or with much reduced persistent laminae, loosely imbricate, brownish, often somewhat glossy, striate, persistent; the upper ones slightly shorter than the internodes, dark green, loose or sometimes somewhat inflated, papery, smooth, glabrous. Ligule 0·7–3 mm. long, obliquely truncate. Leaf-laminae 3–13(–17) × 0·15–0·4(0·6) cm., narrowly linear, tapering to a fine point, expanded or more often involute, fairly stiff, straight, erect or ascending, olive-green, asperulous on the upper surface, glabrous or scattered pilose on the lower surface. Panicle 6–14 × 0·5–1(–rarely 2·5) cm., linear, linear-oblong or rarely narrowly elliptic in outline, with the branches usually somewhat appressed to the rhachis, otherwise as described for *L. hexandra*. Spikelets (3·3)3·6–4·5(–5·5) × 0·9–1(–1·25) mm., pale green or sometimes tinged with purple, oblong-elliptic in outline. Lemma as long as the spikelet, semi-elliptic-oblong in lateral view, cartilaginous, apex produced into a short acute or subacute prolongation; keel and margins scabrous or very shortly ciliolate; flanks shortly hispid or glabrous. Palea slightly shorter than the lemma but of the same consistency, apex acute or tapering into a short acute prolongation; keels scabrous, flanks shortly hispid or glabrous. Anthers 2–3 mm. long, linear. Caryopsis unknown.

Zambia. N: Lake Chila, Abercorn Distr., 1680 m., 26.iii.1955, *Siame* 624 (K; SRGH). S: Kafue Flats, Mazabuka, 6.iii.1962, *Astle* 1402 (K; SRGH).

Also in Katanga, Angola and Tanzania (and Sudan ?). In shallow water at the edge of streams, ponds and lakes, also in dambos and similar moist habitats.

4. **Leersia tisserantii** (A. Chev.) Launert in Senckenb. Biol. **46**: 137, fig. 11–17 (1965). W. D. Clayton, F.T.E.A., Gramineae: 25, fig. 9 no. 8 (1970). TAB. 7, fig. A. Type from West Tropical Africa.
   Oryza tisserantii A. Chev. in Rev. Bot. Appl. Agric. Trop. **12**: 1024 (1932).— Jacques-Félix in Journ. Agr. Trop. Bot. Appl. **2**: 600–619 (1955); op. cit. **6**: 582 (1959).—Tateoka in Bot. Mag. Tokyo, **76**: 165, 171 (1963). Type as above.

A weak, loosely to densely tufted annual or perennial, sometimes producing slender rhizomes. Culms 15–60 cm. tall, erect or geniculately ascending, 2–5-noded, slender, usually branched from near the base, terete, glabrous, but usually hairy to a varying degree at the nodes, smooth. Leaf-sheaths striate, somewhat loose, membranous-papyraceous, often tinged with purple, smooth, glabrous or rarely scattered pilose, auriculate at the mouth. Ligule 1–2·6 mm. long, truncate, often slightly lacerate. Leaf-laminae 3·25–18 × 0·1–0·2(–0·37) cm., narrowly linear, tapering to a fine point, flaccid, expanded or convolute, glabrous. Panicle 8–19 × 2–7·5 cm., usually open and lax; rhachis slender, somewhat furrowed; branches solitary or rarely adnate, filiform, usually devoid of spikelets in the lower ½, erect or obliquely ascending; racemes 2–3(–5) cm. long, bearing 4–8 spikelets. Pedicels 0·4–0·75(–1) mm. long. Spikelets 4–5 mm. long (without caudae), oblong to narrowly elliptic in outline; callus short, glabrous or hispid. Lemma as long as the spikelet, tapering in a cauda 2·5–7·5 mm. in length, chartaceous, often patterned with fine lateral ridges, glabrous or shortly hispid at the flanks; keel with a row of straight or slightly curved cilias of a varying degree of length and toughness; edges usually ciliolate; lateral nerves distinct but rarely prominent. Palea of same consistency as lemma, tapering to a cauda 1·5–2·5 mm. in length, its margins not ciliolate, keel ciliate like the lemma. Anthers 2·4–2·8 mm. long. Caryopsis as described for the genus, 2·5–3 mm. long.

Zambia. N: Abercorn, Kawimbe, 1770 m., 25.ii.1959, *McCallum-Webster* A155 (K; SRGH).
Also in Katanga and West Tropical Africa. In flooded grassland, in small pools on granite rocks, in Elephant and Hippopotamus wallows, etc.

5. **Leersia nematostachya** Launert in Senckenb. Biol. **46**: 136, fig. 22 (1965). TAB. 7 fig. E. Type: Zambia, Mwinilunga Distr., 0·9 km. S. of Matonchi, *Milne-Redhead* 3928 (K).
   Oryza angustifolia C. E. Hubb. in Hook., Ic. Pl. **35**: 3492 (1951).—Pilg. in Engl. & Prantl, Nat. Pflanzenfam. ed. 2, **14d**: 151 (1956).—Tateoka in Bot. Mag. Tokyo, **75**: 418 (1962); op. cit. **76**: 171 (1963). Type as for *Leersia nematostachya*.

A delicate caespitose annual. Culms up to 70 cm. high, 3–4-noded, simple or branched, erect or ascending from a prostrate base, often rooting at the lower nodes, slender, smooth, glabrous. Leaf-sheaths finely striate, obtusely keeled, auricled at the mouth, smooth, glabrous; the lowermost lax and usually longer than the internodes, Ligule 3–7·5 mm. long, lanceolate, slightly lacerate at the apex. Leaf-laminae 10–30 cm. long, filiform, tapering to a setaceous point, convolute, flaccid, straight or slightly curved, smooth, glabrous. Panicle 3–10 cm. long, narrow, erect; rhachis slender, smooth, glabrous; branches 2–5 cm. long, usually simple, solitary, erect and somewhat appressed to the rhachis, bearing 2–6 spikelets. Pedicels c. 1 mm. long. Spikelets 5–8 mm. long (excl. the pseudo-awn), 1–1·3 mm. wide, oblong in outline, green or somewhat tinged with purple. Lemmas as long as the spikelet, narrowly oblong, tapering to a pseudo-awn, chartaceous; lateral nerves distinct; keel shortly ciliate in the upper, scabrous in the lower ½; margins smooth, pseudo-awn 11–18 cm. long, scabrous, usually terete or elliptical from a flattened base, straight, very slender, flexible; callus short, slightly oblique at the base. Palea somewhat shorter and narrower than the lemma but of the same consistency, apex acute; keel shortly ciliate in the upper ⅓, otherwise scabrous. Anthers 3–4 mm. long, linear. Caryopsis narrowly oblong, brownish, 3·5–4·5 mm. long.

Zambia. W: Mwinilunga Distr., 6 km. N. of Kalena Hill, 12.xii.1963, *Robinson* 5897 (K; SRGH).
Also recorded from the Moxico Distr. of Angola. Growing in shallow water in laterite plains, locally dominant.

6. **Leersia oncothrix** C. E. Hubb. in Kew Bull. **1934**: 263 (1935).—Melderis in Svensk Bot. Tidskr. **40**: 228 (1946).—Launert in Senckenb. Biol. **46**: 140, fig. 1, 21 (1965). TAB. 7 fig. D. Type: Zambia, Kalomo, *Trapnell* 993 (K).

A loosely caespitose perennial, often with short rhizomes. Culms 30–85 (–120) cm. high, 2–4-noded, erect or geniculately ascending, often rooting at the lower nodes, simple or branched, subterete, slender, smooth, glabrous but pubescent at the nodes. Leaf-sheaths striate, somewhat loose, smooth, glabrous, auriculated at the mouth; the lowermost usually longer than the internodes and often splitting into irregular fibres. Ligule 2·3–6(–9) mm. long, oblong-lanceolate to triangular-lanceolate, often lacerate at the apex. Leaf-laminae 5–25(–33) × 0·2–0·36 cm., narrowly linear, tapering to a setaceous point, expanded, conduplicate or with the margins involute, usually slightly curved, flexible, glabrous, smooth or scaberulous on both surfaces. Panicle 10–28 × 1·5–3 cm., narrowly elliptic to oblong in outline, erect or somewhat nodding, often purplish; rhachis slender, terete or obtusely angular, asperulous; branches 8–12 cm. long, solitary or rarely adnate, slender, erect or rarely somewhat spreading, branched in the upper half; racemes bearing 3–7(–9) spikelets. Pedicels less than 1mm. long. Spikelets 3·5–4(–4·25) × (1·5)1·7–2 mm., elliptic-oblong or oblong; callus short. Lemma as long as the spikelet, semi-elliptic-oblong in lateral view, cartilaginous, finely transversely ridged; keel and margins with a row of semi-circularly curved cilia; flanks more or less densely covered by inwardly curved hairs similar to the cilia along the margins; lateral nerves distinct. Palea slightly shorter and narrower than the lemma and of the same consistency; keels and flanks as in the lemma; margins glabrous. Anthers c. 2·5 mm. long, linear. Caryopsis unknown.

Zambia. N: Kasama, Chibutubutu flood plain, 1220 m., 19.ii.1962, *Vesey-FitzGerald* 3748 (BM; SRGH). S: Mapanza E., 21.iii.1954, *Robinson* 626 (K; SRGH).

In flooded grassland or small pools, around edges of termite mounds, on sour soil.

7. ORYZA L.

Oryza L., Sp. Pl. **1**: 333 (1753); Gen. Pl. ed. 5: 155 (1754).

Spikelets shortly pedicelled, 1-flowered, solitary, laterally strongly compressed, deciduous or (in cultivated rice) persistent, ☿; rhachilla articulating above the rudimentary glumes.* Glumes 2, reduced to a narrow entire or 2-lobed rim, forming a cup-shaped cavity at the apex of the pedicel, whitish, green or reddish in colour. Sterile lemmas 2, subequal in shape and size, usually very small (nearly as long as the spikelet only in one sp. from S. America), faintly 1-nerved or nerveless. Fertile lemmas 5-nerved (one pair of nerves very close to the margins), cymbiform, strongly keeled, coriaceous or rarely chartaceous, often tessellate or rugulose, usually awned, rarely awnless (in cultivated rice), usually with 2 short lateral protrusions at the apex; awn usually much longer than the spikelet, sometimes coloured, stiff or flexuous, often with a coloured basal callus. Palea slightly shorter than the lemma, of the same consistency and indument but with the margins hyaline, 3- (rarely 5-) nerved, awnless, obtuse or acute to acuminate, usually tightly clasped by the inflexed margins of the lemma. Lodicules 2, small, obovate or lanceolate-elliptic. Stamens 6, with the filaments free. Ovary glabrous. Stigmas pilose, large, laterally exserted. Caryopsis (usually firmly enclosed by lemma and palea thus forming the " husk "), laterally compressed, oblong to oblong-elliptic in outline, whitish to brown or reddish; embryo very small; hilum large, linear. Annual or perennial grasses, growing in a variety of moist habitats. Culms few- to many-noded, erect or ascending, sometimes floating, often spongy, usually hollow between the nodes. Ligule membranous, often lacerate, glabrous or pilose. Inflorescence an open or contracted panicle, many-flowered, erect or drooping.

A genus of 20–25 spp., distributed throughout the tropics and subtropical zones of both hemispheres.

Sterile lemmas subulate or very narrowly triangular. Spikelets 1–1·5(–2) mm. wide
- - - - - - - - - - - - - - - - - 1. *brachyantha*
Sterile lemmas lanceolate or lanceolate-triangular or ovate-lanceolate. Spikelets wider than 2 mm:
  Spikelets 5–6·25 mm. long, transversely attached to the pedicels - - 2. *punctata*
  Spikelets usually more than 7 mm. long, obliquely attached to the pedicels:
    Ligule of the lower leaves usually less than 6 mm. (very rarely up to 10 mm.) long, always truncate or rounded at the apex - - - - - 3. *barthii*
    Ligule of the lower leaves 15–50(60) mm. long, always with the apex acute:
      Stout perennial grass with long creeping branched rhizomes. Spikelets deciduous, always awned - - - - - - - - 4. *longistaminata*
      Annual or short-lived perennial grass without rhizomes. Spikelets persistent, usually awnless but sometimes awned. Cultivated plant - - 5. *sativa*

---

* The spikelet structure of *Oryza* has been differently interpreted by various workers. Here the interpretation by Stapf and Arber has been adopted as in most contemporary floras.

1. **Oryza brachyantha** Chev. & Roehr. in Compt. Rend. Acad. Sci. **159**: 561 (1914); in
Rev. Bot. Appl. **12**: 1022 (1932); op. cit. **14**: 133 (1934).—Bor, Grasses of B.C.I. &
P.: 604 (1960).—Tateoka in Bot. Mag. Tokyo, **75**: 418 (1962); op. cit. **76**: 167, 171
(1963).—Berhaut, Fl. Sénégal: 416 (1967).—W. D. Clayton, in Kew Bull. **21**: 488
(1968). TAB. **8** fig. B.   Type from West Tropical Africa.

A weak caespitose annual.   Culms 30–80(–100) cm. tall, 3–6(–8)-noded, usually
geniculately ascending or prostrate, producing aerial roots at the nodes, striate,
smooth, glabrous.   Leaf-sheaths usually shorter than the internodes, scarious,
loose, striate, smooth, glabrous, shortly auricled and often bearded at the mouth;
the lowermost often bladeless, scale-like, brownish.   Ligule 1–2 (rarely more) mm.
long, truncate, sometimes lacerate.   Leaf-laminae 7–19 × 0·1–0·5 cm., narrowly
linear, tapering to an acute point, green often tinged with purple, flaccid, smooth
or somewhat asperulous on the upper surface mainly along the margins, glabrous,
finely nerved; the midrib distinct.   Panicle 13–30 × 2·5–5 cm., very narrowly
obconical, erect or very rarely nodding, dense; rhachis rather stout, obtusely
angular, smooth, glabrous; branches usually somewhat appressed to the rhachis or
obliquely ascending, angular, smooth or asperulous.   Pedicels 1·5–2·5 mm. long,
robust.   Spikelets 6·5–9·25 × 1·25–1·5 mm., deciduous, obliquely attached to the
pedicel, linear-oblong in lateral view.   Glumes reduced to a short rim which is
entire or inconspicuously 2-lobed.   Sterile lemmas very short, subulate or very
narrowly triangular, acute, always appressed to the fertile floret; the inferior
1·3–2 mm., the superior 1·75–2·5 mm. long.   Fertile lemma slightly shorter than
the spikelet, cymbiform, asymmetrically oblong-linear in outline, chartaceous,
pale-green or straw-coloured, with the nerves usually darker green; flanks glabrous
and smooth or shortly hispid to scaberulous; keel usually softly ciliate mostly
towards the apex; the apical lateral protrusions usually distinct; awn 7–17 cm.
long, very slender, hardly more than 0·5 mm. in diam. at the base, straight or
somewhat wavy in the upper ⅓, scabrous, pale-green to whitish.   Palea about as
long as the lemma, very narrow, similar to the lemma in consistency and indument,
apex usually drawn out into a short point.

   Zambia.   N: Mporokoso Distr., E. side of Mweru-wa-Ntipa, 23 km. SW. of Bulaya,
19.iv.1961, *Phipps & Vesey-FitzGerald* 3294 (BM: K; SRGH).
   Also in Guinea, Sierra Leone, French Sudan and the Congo.   Growing in shallow
ponds, dambos or similar habitats, often locally abundant, but in general a rare species
of wild rice.

2. **Oryza punctata** Kotschy ex Steud., Syn. Pl. Glum. **1**: 3 (1854).—Dur. & Schinz,
Consp. Fl. Afr. **5**: 788 (1894).—Eggeling, Annot. List Grass. Uganda: 29 (1947).—
F. W. Andr. Fl. Pl. Sudan, **3**: 494 (1956).—Bogdan, Rev. List Kenya Grass.: 35
(1958).—Harker & Napper, Ill. Guide Grass. Uganda: 43 (1960).—Tateoka in
Bot. Mag. Tokyo **75**: 418, 419, 422, 423, 425 (1962); op. cit. **76**: 170 (1963); op. cit.
78: 156 (1965).—Napper, Grass. Tangan.: 49 (1965).—W. D. Clayton, F.T.E.A.
Gramineae: 31, fig. 10 no. 1–3 (1970). TAB. **8** fig. A. Type from the Sudan Republic.
   Oryza schweinfurthiana Prodoehl in Bot. Arch. **1**: 231 (1922).—F. W. Andr. Fl.
Pl. Sudan **3**: 494 (1956).   Type from the Congo Republic.
   Oryza sativa var. *punctata* (Kotschy ex Steud.) Kotschy apud Schweinf. in Ber.
Deutsch. Bot. Ges. **44**: 165 (1926).   Type as for *Oryza punctata*.
   Oryza eichingeri var. *longiaristata* Peter in Fl. Deutsch Ost-Afr. **1**: 251, Anh.: 75
(1930).   Type from Tanzania.

A caespitose perennial.   Culms 50–120(–150) cm. tall, 3–5-noded, erect or
geniculately ascending, branched, terete, striate, smooth, glabrous.   Leaf-sheaths
scarious, often spongy and aerenchymatous, distinctly striate, rather loose and often
slipping off the culm, rounded or somewhat keeled in the upper part, auricled at
the mouth, smooth, glabrous.   Ligule 3–10 mm. long, rounded, truncate or somewhat acute, sometimes lacerate, glabrous.   Leaf-laminae 15–45 × 0·5–2·5 cm.,
linear to very narrowly elliptic, acuminate, usually broadest around the middle,
pale-green or rarely glaucous, rather flaccid, expanded or folded around the midrib,
usually asperulous on both surfaces; midrib distinct beneath, some nerves finer
than others.   Panicle 15–35 × 3–17 cm., narrowly to broadly elliptic or sometimes
fan-shaped in outline, loose, erect, or drooping to some extent; rhachis obtusely
angular, glabrous, smooth or inconspicuously scaberulous; branches spreading or
ascending, solitary or sometimes adnate, angular, scabrous; pedicels 2–5 mm.

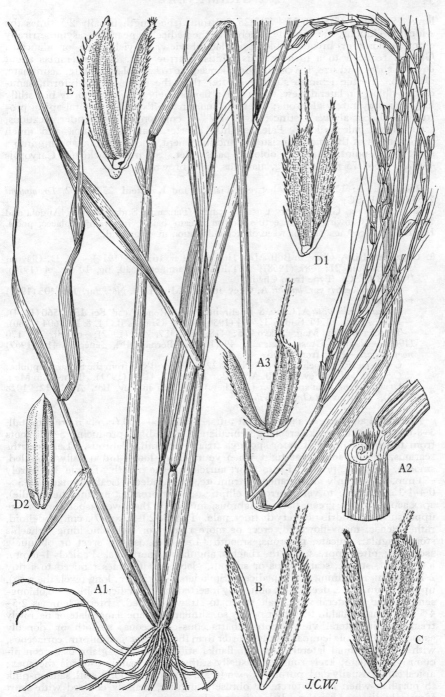

Tab. 8. A.—ORYZA PUNCTATA. A1, habit (×⅔); A2, apex of leaf-sheath and base of leaf-lamina in lateral view, showing auricles and part of ligule (×6); A3, spikelet (×4) all from *Drummond 7754*. B.—ORYZA BRACHYANTHA, spikelet (×4) *Phipps & Vesey-FitzGerald 3294*. C.—ORYZA BARTHII, spikelet (×4) *Burtt 2547*. D.—ORYZA LONGISTAMINATA. D.1 spikelet (×4) *Vesey-FitzGerald 2611*; D2, caryopsis in lateral view (×4) *Wiehe N/494*. E.—ORYZA SATIVA, spikelet (×4) *Chase 5378*.

long. Spikelets (5)5·5–6·25 + (2)2·25–2·8 mm. (the length usually 2·5 times the
width), deciduous, transversely attached to the pedicel or nearly so, asymmetrically
elliptic-oblong or broadly oblong in lateral view, greyish-green or glaucous.
Glumes reduced to a membranous whitish narrow rim. Sterile lemmas about
equal in shape and size, 1–1·5 mm. long, lanceolate to lanceolate-deltate, acuminate,
glabrous. Fertile lemmas slightly shorter than the spikelet, cymbiform, semi-
elliptic-oblong in lateral view, coriaceous; flanks finely tesselate, shortly but stiffly
hispid or very rarely glabrous; keel and margins stiffly ciliate, lateral apical protrusions almost always distinct; awn (1)2–7·5 cm. long, very slender, flexuous,
scaberulous, pale yellow. Palea slightly shorter than the lemma and much
narrower but of the same consistency and indument, apex acute or tapering into a
short acute point. Anthers oblong, pale-violet. Stigmas blackish. Caryopsis
4–4·75 × 1·5–1·75 mm., oblong, glabrous, light brown.

Rhodesia. S: Urumbo Pan between Fishans and Kapateni, 25.iv.1962, *Drummond*
7754 (BM; LISC; SRGH).
Also in Nigeria, Ghana, Ivory Coast, Uganda, Tanzania, Sudan, Congo, Angola and
Madagascar. Growing in dense tussocks on stream banks, in swampy places, pools.
Has recently become a serious weed of rice-cultivation in Zwaziland.

3. **Oryza barthii** A. Chev. in Bull. Mus. Hist. Nat. Par. **16**: 405 (1911).—W. D. Clayton
in Kew Bull. **21**: 487 (1968); F.T.E.A., Gramineae: 30, fig. 10 no. 4 (1970).
TAB. **8** fig. C. Type from Chad.
    Oryza sylvestris var. *barthii* A. Chev. in Bull. Mus. Hist. Nat. Par. **16**: 405 (1911)
nom. nud.
    Oryza breviligulata A. Chev. & Roehr. in Compt. Rend. Acad. Sci. **159**: 560 (1914).
—F. W. Andr., Fl. Pl. Sudan **3**: 494 (1956).—Bor, Grasses B.C.I. & P.: 604 (1960).
—Tateoka in Bot. Mag. Tokyo, **75**: 418, 455, 457, 460 (1962);—op cit. **76**: 170
(1963).—Napper, Grass. Tangan.: 49 (1965).—Berhaut, Fl. Sénégal: 416 (1967)
nom. illegit.—Type from Chad.
    Oryza mezii Prodoehl in Bot. Arch. **1**: 223 (1922). Type from the Sudan Republic.
    Oryza stapfii Roshev. in Bull. Appl. Bot. Pl. Br. **27**, 4: 51 (1931). Type from Mali.
    Oryza perennis subsp. *barthii* (A. Chev.) A. Chev. in Rev. Bot. Appl. **12**: 1028
(1932). Type as for *Oryza barthii*.

An annual aquatic grass growing in tufts. Culms 60–120 (rarely more) cm. tall,
3–8-noded, rather weak, erect or geniculately ascending, producing aerial roots
from the lower nodes, terete, spongy, striate, smooth, glabrous. Leaf-sheaths
scarious, striate, somewhat tight when young later loose and usually wrinkled,
smooth, glabrous, produced into short auricles at the mouth. Ligule 3–6 (rarely
–9) mm. long, thinly membraneous, truncate, or rounded. Leaf-laminae 15–45 ×
0·4–1·3 cm., linear to very narrowly elliptic (always broadest around the middle),
apex acute, intense green, flaccid, glabrous, smooth on the lower, asperulous on the
upper surface; midrib not very distinct, pale. Panicle 20–35 × 3–7·5 cm., obdeltoid,
rather dense, many-flowered, erect or more rarely somewhat nodding; rhachis
stout, angular, sulcate, glabrous, smooth; branches usually erect or obliquely
ascending, often appressed to the rhachis, angular, scaberulous. Pedicels 1–6 mm.
long, stout, striate, scaberulous or smooth, glabrous. Glumes reduced to a tiny
2-lobed rim, remaining at the pedicel. Spikelets 8–10·5 mm. long (excl. the awn),
up to 3·4 mm. wide, deciduous, obliquely inserted on the pedicel, oblong to oblong-
semi-elliptic in lateral view, pale green to straw coloured. Sterile lemmas 2·5–
4·5 mm. long, equal in length or nearly so, similar in shape, lanceolate or narrowly
triangular in lateral view, acute, chartaceous-coriaceous, smooth or dorsally
asperulous. Fertile lemma slightly shorter than the spikelet, cymbiform, coriaceous,
with 2 longitudinal lateral grooves; flanks stiffly hispid to glabrous, inconspicuously tessellate; keels rounded, usually stiffly ciliate mainly towards the apex;
apical callus usually dark purplish; awn (6·5)8–16(–19) cm. long, stiff, salmon-pink
to purplish when fresh, terete or obtusely angular, densely covered with short
forwardly directed bristles. Palea about as long as the lemma but much narrower,
similar in texture and indument, with the apex drawn out in short blunt purplish
point.

Zambia. B: Mongu, Barotse flood plain, 19.iii.1964, *Verboom* 1152 (SRGH). S:
Musialoulozi, Kalumbu, on Cordon road, Kafue National Park, Namwala, 30.i.1962,

Mitchell 12/79 (BR; K; LISC; SRGH). **Rhodesia.** N: Urungwe, Mana Pools, Zambezi Valley, 18.iv.1967, *Cleghorn* 1544 (BM; SRGH).

Known from Mauritania, Senegal, Guinea, Sierra Leone, Liberia, Ivory Coast, Sudan Republic, Ghana, Nigeria, Sudan, Tanzania and Uganda. Growing in shallow ponds, vleis, dambos, rice fields and similar habitats, often forming pure dominant stands but usually scattered with other aquatic grasses.

4. **Oryza longistaminata** A. Chev. & Roehr. in Compt. Rend. Acad. Sci. **159**: 561 (1914). —A. Chev., Expl. Bot. Afr. Occ. Fr. **1**: 739 (1920).—W. D. Clayton in Kew Bull. **21**: 488 (1968); F.T.E.A., Gramineae: 30, fig. 10 no. 6, 8 (1970).—Launert in Merxm., Prodr. Fl. SW. Afr. 160: 134 (1970). TAB. **8** fig. D. Type from Chad.

Oryza sylvestris A. Chev. in Bull. Mus. Hist. Nat. Par. **16**: 405 (1911) *nom. nud.*

Oryza barthii sensu Eggeling, Annot. List Grass. Uganda: 29 (1947).—sensu Sturgeon, Rev. List Grass. S. Rhod. in Rhod. Agric. Journ. **51**: 213 (1954).—sensu Chippindall in Meredith, Grass. & Pastures of S. Afr.: 32 (1955).—sensu Pilg. in Engl. & Prantl, Nat. Pflanzenfam. ed. 2, **14d**: 143, 144, 145, 149 (1956).—sensu Troupin, Fl. Garamba, **1**: 60 (1956).—sensu F. W. Andr., Fl. Pl. Sudan **3**: 493 (1956).—sensu Jackson & Wiehe, Annot. Check List Nyasal. Grass.: 50 (1958).—sensu Bogdan, Rev. List Kenya Grass.: 35 (1958).—sensu Harker & Napper, Ill. Guide Grass. Uganda: 43 (1960).—sensu Tateoka in Bot. Mag. Tokyo, **75**: 418, 455, 456, 457, 460 (1962).—sensu Compère in Bull. Jard. Bot. Brux. **33**, 3: 397 (1963).—sensu Berhaut, Fl. Sénégal: 416 (1967) non. A. Chev. (1911).

Oryza perennis sensu Wild in Pl. Aquat. Nuis. Afr. et Madag. in C. S. A. Bull. **14**: 21 (1964).—sensu Napper, Grass. Tangan.: 49 (1965) non Moench.

A robust perennial with long creeping branched rhizomes. Culms up to 250 cm. tall or more, up to 2·5 cm. or more in diam., (2)4–10-noded, erect or ascending, sometimes floating, weak, spongy, the submerged parts whitish, developing aerial roots from the nodes, glabrous, smooth. Leaf-sheaths scarious, spongy, usually tessellate, pale-green to brownish, shorter or nearly as long as the internodes, with up to 15 mm. long narrow auricles at the mouth, glabrous, smooth. Ligule (0·8)1·5–5·5 cm. long, narrowly triangular, acute. Leaf-laminae 10–75 × 0·5– 2·5 cm., linear, linear-lanceolate to very narrowly elliptic, broadest below the middle, acuminate, bright to dark green, somewhat flaccid, asperulous on the upper, smooth or somewhat asperulous on the lower surface, usually scabrous along the margins, glabrous, midrib often indistinct. Panicle 16–40 × 2·5–8 cm., narrowly oblong to narrowly elliptic in outline, erect or slightly drooping, dense, many-flowered; rhachis obtusely angular, usually smooth, glabrous; branches obliquely ascending to nearly erect, somewhat stiff, angular, scaberulous, glabrous but usually with a tuft of short hairs at the very base. Pedicels 0·5–4(–7) mm. long, stout, angular, scabrous or smooth. Spikelets 7–12(–15) mm. long (excl. awn) and 2–3 mm. wide, deciduous, obliquely attached to the pedicel, asymmetrically elliptic-oblong in lateral view, pale-green to brownish. Glumes reduced to a narrow membranous rim. Sterile lemmas nearly equal in shape and size, (2)2·5–3·8(–4·5) mm. long, lanceolate to broadly lanceolate, acute to acuminate, sometimes 3-dentate with the central tooth dominating the lateral (see fig. D.1), glabrous, smooth. Fertile lemma slightly shorter than the spikelet, cymbiform, semi-elliptic-oblong, coriaceous; flanks finely tessellate, stiffly hispid to a varying degree; keels usually stiffly ciliate towards the apex; lateral apical protrusions mostly distinct; awn (2·6)4–7·5 cm. long, rather slender, straight or slightly wavy, usually pink when fresh (at least in the upper part) or often purplish, scabrous with short stiff forward directed bristles. Palea slightly shorter than the lemma and much narrower but of the same consistency and indument; apex acute or tapering into a subacute point, somewhat outwardly curved. Stamens 6; anthers 4·5–5·5 mm. long, linear-oblong. Stigmas blackish. Caryopsis 7·5–8·5 mm. long, oblong in lateral view, glabrous, light-brown, glossy.

Botswana. N: Ngamiland, Matlapaneng, 940 m., iv.1958, *Robertson* 668 (BM; SRGH). **Zambia.** B: Barotse plain, 980 m., vi.1933, *Trapnell* 1343 (K). N: Abercorn Distr., edge of Kawimbe Dambo, 1710 m., 26.iii.1959, *McCallum Webster* A241 (K; LISC; SRGH). C: Broken Hill, Lukangawaya landing, 12.x.1962, *Vesey-FitzGerald* 3756 (BM; SRGH). E: Fort Jameson, iii.1962, *Verboom* 594 (K). S: Kalumba on Cordon road, Kafue National Park, 30.i.1962, *Mitchell* 12/80 (BR; K; LISC; SRGH). **Rhodesia.** N: Gokwe Distr., Lutope vlei, 18.iii.1963, *Bingham* 533 (BM; LISC; SRGH). W: Shangani Distr., Gwampa Forest Res., ii.1954, *Goldsmith* 21/54 (K; SRGH). C: Hartley, Poole Farm, 30.iii.1946, *Hornby* 2442 (K; SRGH). **Malawi.** N: Karonga

Distr., Rukuru R., 28.iv.1937, *Smaltey* (K). S: Flood plains of Lake Chilwa, 28.i.1950, *Wiehe* N/494 (K). **Mozambique.** N: Mogincual, 27·2 km. from Liupo towards Mogincual, 30.iii.1964, *Torre & Paiva* 11493 (LISC). Z: Namacurra-Quelimane, 25.iii.1943, *Torre* 4973 (LISC). SS: Caniçado Guija, 11.v.1944, *Torre* (LISC).

The species is confined to Africa where it is recorded from Senegal, Mali, Sierra Leone, Guinea, Liberia, Ivory Coast, Ghana, Nigeria, Cameroon, Sudan, Gabon, Congo, Uganda, Kenya, Tanzania, Angola, northern Transvaal, SW. Africa and Madagascar. In shallow water of pans, pools, in stagnant or running swamps, flood plains etc., also edges of rivers, dams or on river banks, sometimes also amongst cultivated rice; known as hippopotamus fodder, often occurring in pure stands, sometimes mixed with *Vossia*.

5. **Oryza sativa** L., Sp. Pl. **1**: 333 (1753).—Engl., Pflanzenw. Ost-Afr. **B**: 59, 64 (1895); op. cit., **C**: 106 (1895).—Rendle in Cat. Afr. Pl. Welw. **2**, 1: 231 (1899).—Monro in Proc. Rhod. Sci. Ass. **6**, 1: 60 (1906).—R. E. Fr., Wiss. Ergebn. Schwed. Rhod.-Kong. Exped. **1**: 205 (1916).—Stent in Bothalia, **1**, 4: 275 (1924).—Sturgeon, Rev. List. Grass. S. Rhod. in Rhod. Agric. Journ. **51**: 213 (1954).—Chippindall in Meredith, Grass. & Pastures of S. Afr.: 32 (1955).—Jackson & Wiehe, Annot. Check List Nyasal. Grass.: 68 (1958).—Tateoka in Bot. Mag. Tokyo, **75**: 418, 455, 456, 460 (1962); op. cit. **76**: 168 (1963).—Berhaut, Fl. Sénégal: 416 (1967).—W. D. Clayton in F.T.E.A., Gramineae: 28, fig. 10 no. 7 (1970). TAB. **8** fig. E. Type from India?

For a complete synonomy of this cultivated species see Tateoka in Bot. Mag. Tokyo **76**: 168 (1963): for infraspecific classification see Portères in Journ. Agric. Trop. **3**: 341, 541, 627, 821 (1956).

An annual or rarely short-lived perennial of variable habit. Culms 45–180 cm. tall, 3–many-noded (up to 20), 6–8 mm. in diam., erect or ascending from a geniculate base, terete, hollow, smooth, glabrous. Leaf-sheath coarsely striate, tight when young, later somewhat loose, often somewhat spongy, green or sometimes tinged with brown or purple, smooth, glabrous; the lowest usually longer, the upper shorter than the internodes. Ligule (1·25)1·5–3 cm. long, triangular, acute, entire or split, usually glabrous, sometimes tinged with pink, purple or brown in some varieties or strains. Leaf-laminae 12–65 × 0·4–1·75 cm., linear, tapering to an acute point, bright green to glaucous, rather flaccid, glabrous or puberulous, rarely scattered with short hairs, smooth on the lower, asperulous on the upper surface; midrib usually distinct. Panicle up to 50 cm. long, erect, curved or drooping, very variable in density; rhachis obtusely angular, smooth, glabrous or scattered hairy; branches solitary or clustered, forming a variable angle with the rhachis (from being nearly erect to spreading), angular, scabrous. Pedicels up to 4 mm. long, stout, scabrous. Spikelets 8–11 × 2·5–3·5 mm., not deciduous (the articulation to the pedicel is usually completely solidified but there are, however, some strains in which the spikelets fall off at maturity), obliquely attached to the pedicel, very variable in shape, asymmetrically oblong to elliptic-oblong in lateral view. Glumes (the narrow rim at the base of the spikelet) varying in colour from pale white to yellow, purple or black. Sterile lemmas about equal in shape and size, usually 2–3 mm. long (very rarely up to ⅓ the length of the spikelet in certain forms "winged varieties"), lanceolate, acute, glossy. Fertile lemma as described for *O. barthii*, sometimes coloured, usually awnless, sometimes awned (the awn often purple-pink, more rarely colourless). Paleas as described for *O. barthii*. Sexual organs and caryopsis similar to those of *O. barthii*.

Mozambique. N: Mogovolas, Posto Agrícola, 9.vi.1934, *Ribeiro* (LISC).
Cultivated in the tropics and subtropics throughout the world.

V. PHYLLORHACHIDEAE C. E. Hubb.

Phyllorhachideae C. E. Hubb. in Hook., Ic. Pl. **34**: 3386, 5 (1939).

Inflorescence spike-like, terminal (sometimes with reduced axillary inflorescences within the upper leaf-sheaths), consisting of short racemes inserted along a leaf-like rhachis; the membranous margins of the rhachis spathe-like, enclosing the inflorescence; racemes with a flattened axis, falling entire, with 1–4 spikelets. Spikelets unisexual, with the sexes within the same inflorescences or in separate inflorescences, 2-flowered, awnless, ♀ spikelets larger than the ♂; rhachilla sometimes extended. Glumes 2, shorter than the spikelet. Inferior floret reduced to a coriaceous lemma. Superior floret: lemma coriaceous, partly embraced by the inferior lemma, with 5 or more nerves; palea 2–12-nerved; lodicules 2; vestigial ovary usually present; the female spikelet with 3–6 staminodes. Caryopsis with a linear hilum; embryo small, bambusoid; starch grains compound. Basic chromosome number 12.

A tribe of forest grasses with bamboo-like culms. Leaf-laminae with pseudopetioles and inconspicuous tessellate venation. Genera 2, in Africa and Madagascar.

8. PHYLLORHACHIS Trimen

Phyllorhachis Trimen in Journ. of Bot. **17**: 353 (1879).

Inflorescence as described for the tribe; axillary inflorescences all ♀; racemes short, usually not more than 1–2; spikelets longer than the ones of the terminal inflorescences, with the elongated stigmas protruding from the subtending leaf-sheaths, the terminal one ☿; the lowermost spikelet of each raceme ♀, the following ♂, sometimes reduced spikelets present; rhachilla not extended. ♂ spikelets; inferior glume rather narrow, almost subulate, acute; the superior 1-nerved, ovate-oblong, up to ⅔ the length of the spikelet; lemma of the inferior floret 3-nerved, coriaceous; lemma of the superior floret 5–7-nerved, thinly coriaceous, keeled, its palea 2-nerved; stamens 6. ♀ spikelets: inferior glume subulate or even setaceous; the superior narrowly ovate to oblong, up to ½ the length of the spikelet, 5–9-nerved; lemma of the inferior floret embracing the superior floret, dorsally with a longitudinal furrow, rather rigid, many-nerved; lemma of the superior floret 11–many-nerved, membranous to thinly coriaceous, its palea 8–12-nerved. Caryopsis free, elliptic-oblong in outline, with a rather shallow longitudinal furrow; embryo minute; hilum linear, almost as long as the caryopsis. Perennial grasses, sometimes of shrubby habit, with branched slender culms. Ligule small, ciliate.

A monospecific genus confined to east tropical and south tropical Africa.

Phyllorhachis sagittata Trimen in Journ. of Bot. **17**: 205, 355 (1879).—Stapf & Hubb. in F.T.A. **9**: 1089 (1934).—Pilg. in Notizbl. Bot. Gart. Berl. **12**: 701 (1935).— C. E. Hubbard in Hook., Ic. Pl. **34**: 3386 (1939).—Jackson, Annot. Check-List Nyasal. Grass.: 54 (1958).—Napper, Grass. Tangan.: 50 (1965).—C. E. Hubbard apud W. D. Clayton in F.T.E.A. Gramineae: 34, fig. 12 (1970). TAB. **9**. Type from Angola.

A loosely caespitose perennial (but frequently flowering in the first year), often of a shrubby habit. Culms up to 130 cm. tall, decumbent, rarely erect, terete, many-noded, branched, retrorsely scabrid or sometimes smooth; nodes stiffly retrorse-pilose to pubescent. Leaves evenly distributed over the culm. Leaf-sheaths coarsely striate, usually tight, sometimes slipping off the culm, retrorsely hispid to scabrid. Leaf-laminae slightly asymmetric, 3·5–12 × 1·5–2·2 (2·7) cm., lanceolate, narrowly elliptic to oblong-lanceolate, base sagittate, apex acute to acuminate, scabrid, sometimes with scattered stiff hairs along the nerves, margins often ciliate; auricles up to 12 mm. long, acute or subobtuse; pseudo-periole 0·5–2·5 mm. long. Terminal inflorescence (3)4·5–12 cm. long, erect, straight or slightly curved, pale green; main axis flattened, 3–9 mm. wide; individual racemes up to 20 in number, 6–10 mm. long, with the rhachis c. 2·5 mm. wide. ♂ spikelets 5·5–8 mm. long, lanceolate to lanceolate-oblong in outline. ♀ spikelets narrowly ovate, slightly asymmetric, up to 16 mm. long, often scabrid. Spikelets of the axillary inflorescences up to 25 mm. long, their stigmas up to 25 mm. long or more.

Zambia. B: Mankoya, Lalefuta R. bank, 3.iv.1964, *Verboom* 1170 (SRGH). N: Mpika, M'fuwe Camp in Chilongozi, 24.iii.1963, *Verboom* 921 (BM; SRGH). W: Ndola Fire Protection Plots, 1250 m., 1.v.1953, *Hinds* 139 (SRGH). C: Katondwe, 4.iv.1966, *Fanshawe* 9661 (SRGH). E: Fort Jameson, Rukuzi R./Lundazi road, 24.xii. 1963, *Wilson* 8 (SRGH). S: Kafue National Park, Luansanda R. ford, 29.iv.1962, *Uys* 23 (BM; SRGH). **Malawi.** C: Fort Manning, banks of Livulezi R., 12.x.1951, *Jackson* 621 (K; SRGH). S: Cholo, Tung Station, 24.xi.1950, *Jackson* 298 (K; SRGH). **Mozambique.** MS: Chimoio, R. Vanduzi, 610 m., 4.iv.1952, *Chase* 4433 (BM; LISC; SRGH).

Also known from Tanzania and Angola. In riverine vegetation in sandy soil.

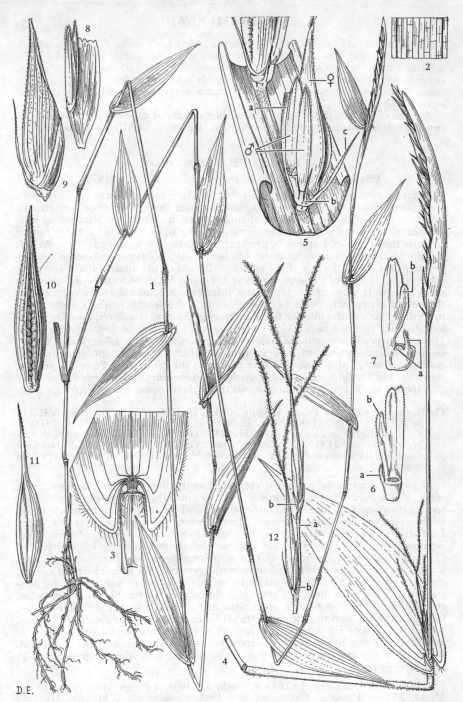

Tab. 9. PHYLLORHACHIS SAGITTATA. 1, habit (×½) *Astle* 4427 and *Mitchell* 2635; 2, detail from underside of leaf-lamina showing venation (×5); 3, base of leaf-lamina (×1½); 2–3 from *Chase* 7523; 4, branch with axillary and terminal inflorescences (×1) *Astle* 4427; 5, raceme showing ♂ and ♀ spikelets, secondary rhachis (a), vestigial spikelet (b) and inferior glume of ♀ spikelet (c) (×3½); 6, 7, secondary rhachis, front and dorsal views, showing vestigial spikelets at (a) and (b) (×9); 8, ♂ spikelet on secondary rhachis (×3½); 9, ♀ spikelet (×3½); 10, dorsal view of inferior lemma from ♀ spikelet (×3½); 11, superior floret of ♀ spikelet (×3½), from *Verboom* 921; 12, axillary inflorescence, showing primary (a) and secondary (b) rhachis (×1½) *Jackson* 1793. From F.T.E.A.

VI. EHRHARTEAE Nevski

Ehrharteae Nevski in Act. Inst. Bot. Acad. Sci. U.R.S.S., **4**: 227 (1937).

Inflorescence usually a panicle, open or contracted. Spikelets strongly laterally compressed, 3-flowered, with the first and second floret reduced to empty lemmas, the third ⚥; rhachilla not extended beyond the uppermost floret, disarticulating above the glumes but persistent between the florets. Glumes 2, equal or unequal, usually shorter than the spikelet, persistent. Sterile lemmas usually equal or nearly so, exceeding the fertile one, sometimes awned, often with basal appendages; fertile lemma smaller and usually thinner in texture. Palea 2-nerved, with the 2 keels close together (rarely 1-nerved and 1-keeled), slightly shorter than the lemma. Lodicules 2, often 2-lobed. Stamens 6, but sometimes 4, 3 or 1. Styles 2; stigmas plumose. Caryopsis laterally compressed; embryo small; hilum linear almost extending over the whole length of the caryopsis; starch grains compound. Chromosomes small, basic number 12.

An Old World tribe of 3 genera, mainly in warm temperate regions of South Africa and Australia.

9. EHRHARTA Thunb.

Ehrharta Thunb., Vet. Akad. Handl. Stockh. **40**: 217, pl. 8 (1779).

Spikelets strongly laterally compressed, solitary, pedicelled, 3-flowered; the first and second florets sterile, reduced to empty lemmas, the uppermost ⚥. Glumes 2, equal or unequal, usually shorter than, rarely as long as, seldom longer than the rest of the spikelet, 1–11-nerved, persistent, membranous, obtusely keeled. Lemmas heteromorphous; the sterile ones (1 and 2) equal or unequal, almost always exceeding the glumes, compressed, usually cartilaginous, awned or muticous, 3–5-nerved, transversely rugose, tuberculate or smooth, the second one often with variously shaped callous appendages at the base; the fertile one (the uppermost) shorter than the sterile ones, often hinged with the second sterile one by basal callous knob-like appendages, usually 5-nerved, indurate but less so than the sterile ones, muticous, sometimes transversely rugose or tuberculate but more often smooth. Paleas slightly shorter or as long as the lemma, 2-nerved, with the nerves very close together, 2-keeled, membranous. Lodicules 2, often lobed. Stamens 6 or 3, very rarely 1. Ovary glabrous, ovoid; styles distinct; stigmas plumose, laterally exserted above the base. Caryopsis elliptic in outline, strongly laterally compressed, embryo c. $\frac{1}{5}$ the length of the caryopsis. Annuals or perennials of varying habit. Leaves with the laminas often greatly reduced. Ligule membranous or scarious but sometimes reduced to a rim of hairs. Inflorescence a panicle, rarely a raceme, usually rather dense, compact or interrupted, sometimes spike-like, more rarely loose and open, sometimes secund.

An African genus of c. 28 spp., mainly concentrated in the south-western districts of the Cape Prov., with only a few spp. in SW. Africa and 1 sp. extending north as far as Ethiopia. A few spp. have been introduced into America, Asia, Australia, New Zealand and Southern Europe.

Ehrharta erecta Lam., Encycl. Méth. Bot. **2**: 347 (1786).—Stapf in Harv. & Sond., F. C. **7**: 662, 671 (1900).—Bor, Grass. B. C. I. & P.: 484 (1960).—W. D. Clayton, F.T.E.A., Gramineae: 38, fig. 14 (1970). TAB. **10**. Type from S. Africa (Cape Prov.).

    Ehrharta abyssinica Hochst. in Flora, **38**: 193 (1855).—Dur. & Schinz, Cosp. Fl. Afr. **5**: 790 (1895).—Eggeling, Annot. List Grass. Uganda: 16 (1947).—Sturgeon in Rhod. Agric. Journ. **51**: 214 (1954).—Robyns & Tournay, Fl. Parc Nat. Alb. **3**: 204 (1955).—Jackson & Wiehe, Annot. Check List Nyasal. Grass.: 37 (1958).— Bogdan, Rev. List Kenya Grass.: 35 (1958).—Harker & Napper, Ill. Guide Grass. Uganda: 27, pl. 78 (1960).—Bor, Grass. B.C.I. & P.: 484 (1960).—Napper, Grass. Tangan.: 50 (1965). Type from Ethiopia.

    Ehrharta erecta var. *abyssinica* (Hochst.) Pilg. in Notizbl. Bot. Gart. Berl. **9**: 508 (1926). Type as above.

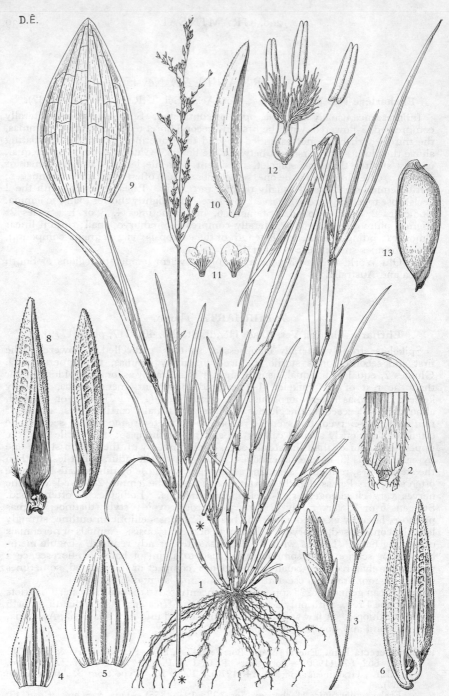

Tab. 10. EHRHARTA ERECTA. 1, habit (× ½); 2, junction of leaf-sheaths and lamina
showing ligule (× 3); 3, branchlets of inflorescence exhibiting (a) spikelet in lateral
view (b) glumes (× 5); 4, inferior glume (× 10); 5, superior glume (× 10); 6, florets
removed from the glumes (× 9); 7, first (sterile) lemma (× 10); 8, second (sterile)
lemma (× 10); 9, third (fertile) lemma (× 10); 10, palea (× 10); 11, pair of lodicules
(× 10); 12, sexual organs (× 10) all from *Fisher & Schweickerdt* 315; 13, caryopsis
in lateral view (× 10) *E.M. & W.* 137.

Ehrharta erecta var. *natalensis* Stapf in Harv. & Sond., F.C. **7**: 671 (1900).—Wood, Natal Pl. **5**, 2: pl. 446 (1905).—Chippindall in Meredith, Grasses & Pastures of S. Afr.: 40 (1955). Type from Natal.

Ehrharta panicea Sm., Pl. Ic. Hact. Ined.: t. 9 (1789). Type from S. Africa (Cape Prov.).

Ehrharta paniciformis Nees ex Trin. in Mém. Acad. Sci. Petersb., Sér. 6, **5**: 64 (1839). Type from S. Africa.

Panicum deflexum Guss. ex Ten., Fl. Nap. **5**: 320 (1835–36). Type from southern Europe.

A rather variable caespitose perennial. Culms 40–100 cm. tall, 4–6-noded, usually geniculately ascending from a decumbent base, often rooting from the lower nodes, branched from below, slender, rather weak, smooth, glabrous or scattered pilose. Leaf-sheaths tight at first, later slipping off the culm, shorter than the internodes, striate, smooth, glabrous or scattered pilose. Ligule 4–7 mm. long, obtuse or truncate, often lacerate. Leaf-laminae 4–20 × 0·2–1(1·5) cm., very narrowly lanceolate to linear, auricled at the base, tapering to a fine soft point, usually expanded, rather flaccid, glabrous or very rarely scattered pilose, smooth or asperulous. Panicle 6–20 cm. long, erect, rarely nodding, open or contracted, irregularly elliptic to almost linear in outline, somewhat stiff; rhachis terete below, obtusely angular above, smooth or very rarely scaberulous; branches distant, very unequal, 2–3-nate or solitary, simple or sparingly branched, usually ascending to almost appressed to the rhachis, rarely spreading, flexuous, filiform, smooth or rarely scaberulous towards the apex. Pedicels up to 1 cm. long, capillary. Spikelets (3)4–5·75(–6·8) mm. long, light green, oblong in outline. Glumes unequal; the inferior 3–3·6 mm. long, 3-nerved but sometimes with short additional nerves on either side close to the margins, ovate-lanceolate, obtusely keeled, apex obtuse to rarely apiculate; the superior 3·5–4·5 mm. long, ovate to ovate-elliptic, 5-nerved, apex subobtuse to acute. Sterile lemmas (1 and 2) slightly unequal, muticous, with the flanks corrugated to a varying degree to completely smooth, sometimes scabrous or very slightly hispidulous, faintly 5-nerved; the inferior c. 5·4 mm. long, with the apex subacute in profile; the superior c. 6·5 mm. long, with a conspicuous basal hinge-like appendage, apex subobtuse in profile. Fertile lemma (the 3rd.) c. 5·5 mm. long, 5-nerved (sometimes with 1 short additional nerve on either side close to the margin), often cross-veined, ovate-elliptic, with the apex obtuse to subacute, keel scaberulous. Palea deeply cymbiform, scaberulous along the keels. Lodicules very small, usually broadly lobed, glabrous. Stamens 6; anthers 1·5–2 mm. long. Caryopsis c. 2 mm. long.

Rhodesia. E: Central Patrol, Stapleford Forest Reserve, 1785–1800 m., 13.vi.1934, *Gilliland* 304 (BM; K; SRGH). **Malawi.** N: Vipya, Mzimba, 29.ix.1950, *Jackson* 181 (K; SRGH). C: Mt. Dedza, 6.ix.1950, *Jackson* 165 (K). **Mozambique.** MS: Manica, Mt. Zuira, Tsetsera, near the residence of Sr. Carvalho, 1980 m., 11.xi.1965, *Torre & Pereira* 12890 (LISC).

In the Cape Prov., Transvaal, Orange Free State, Natal and Lesotho, extending northwards to the Congo, Uganda, Kenya, Tanzania, Ethiopia and the Mascarene Is. Also introduced into southern Europe, Asia and North America. Mainly in forest shade in places with open canopies or in clearings, but also in disturbed areas and on hillsides, occasionally even on coastal sands.

In contrast to all other spp. of the genus, *E. erecta* is adapted to a wide range of edaphic conditions which explains the great variability of the sp. The distinguishing characters between *E. erecta* and *E. abyssinica* given by various workers in the past break down completely if one tries to apply them to the large number of specimens which are now available from all parts of Africa. O. Stapf (loc. cit.) had already regarded the East African plants as being " a slightly different form " of *E. erecta*.

VII. POEAE

Inflorescence a panicle, sometimes a raceme, rarely a spike. Spikelets 2–many-flowered, usually all alike, laterally compressed, with the uppermost floret(s) reduced. Rhachilla disarticulating above the glumes and between the florets, rarely persistent and then the spikelet breaking off as an entire unit. Glumes usually 2 (rarely the inferior completely absent), equal or unequal, persistent, shorter than the rest of the spikelet. Lemmas (3)5–13-nerved, herbaceous or chartaceous, not awned or awned from the entire or slightly 2-dentate apex; awn simple, straight or curved. Paleas as long as or slightly shorter than the corresponding lemmas, 2-nerved, 2-keeled, hyaline. Lodicules 2 or rarely absent. Stamens 3, rarely reduced to 1. Caryopsis with a small embryo; hilum linear or punctiform; starch grains compound. Basic chromosome number 7.

A tribe of over 40 genera, distributed throughout the temperate regions of both hemispheres; a few genera extending into the subtropics and mountain areas of the tropics.

Inflorescence a spike, with the spikelets arranged in 2 rows on alternate sides of the
    rhachis - - - - - - - - - - - - 10. **Lolium**
Inflorescence a panicle:
    Lemmas very broad from a distinctly cordate base (see TAB. **12**, fig. A5), not awned;
        spikelets 3–14 mm. wide; annual plants - - - - 11. **Briza**
    Lemmas lanceolate or ovate-lanceolate, never cordate at the base; spikelets usually
        smaller; perennial plants (except for *Poa annua*):
        Leaf-laminae with transverse veinlets - - - - - 13. **Pseudobromus**
        Leaf-laminae without transverse veinlets:
            Lemmas keeled, laterally compressed, not awned - - - - 12. **Poa**
            Lemmas dorsally rounded (rarely slightly keeled towards the apex), awned or
                awnless - - - - - - - - - - - 14. **Festuca**

10. LOLIUM L.

Lolium L., Sp. Pl. **1**: 83 (1753); Gen. Pl. ed. 5: 36 (1754).

Spikelets sessile, solitary, 3–11-flowered, usually laterally compressed, rarely subcylindrical, awned or muticous; florets ⚥ but the uppermost often reduced; rhachilla disarticulating above the glumes and between the florets, glabrous. Glumes 2 in the uppermost terminal spikelet, 1 (the superior) in the lateral spikelets, persistent, abaxial, 7–9-nerved, linear to oblong, chartaceous to coriaceous, not keeled, with the apex obtuse to acute, awnless. Lemmas 5–7(–9)-nerved, membranous to chartaceous, dorsally rounded, oblong with the apex subobtuse and usually shortly 2-dentate, glabrous, awnless or awned from near the apex. Paleas about as long as the corresponding lemmas, 2-keeled, chartaceous-membranous, elliptic or oblong, apex obtuse and often minutely 2-dentate, keels crested or ciliate. Lodicules 2, minute, ovate-lanceolate. Ovary glabrous, obovate-truncate, styles distinct, rather short, plumose for the whole length. Caryopsis elliptic-oblong to linear-oblong in outline, dorsally with a longitudinal groove, ventrally sulcate, with a short apical appendage; hilum linear. Annuals or perennials of various habitats. Ligule a hyaline membrane. Inflorescence a solitary terminal spike, with the spikelets arranged in 2 ranks on opposite sides of the rhachis and inserted alternately in depressions edgewise to it.

A genus of 6 spp. distributed throughout Europe, North Africa and the temperate regions of Asia; introduced into almost all other parts of the world.

Superior glume (lateral spikelets) as long as or longer than the rest of the spikelet; plant
    a robust annual - - - - - - - - - - 1. *temulentum*
Superior glume (lateral spikelets) always shorter than the rest of the spikelet; plants
    annual, biennial or perennial:
    Lemmas awned to a varying degree; plant annual or biennial; leaf-lamina involute
        when young - - - - - - - - - 2. *multiflorum*
    Lemmas muticous, subobtuse to somewhat pointed; plant perennial (rarely flowering
        in the first year); leaf-lamina plicate when young - - - - 3. *perenne*

1. **Lolium temulentum** L., Sp. Pl. **1**: 83 (1753).—Stapf in Harv. & Sond., F. C. **7**: 738 (1900).—Monro in Proc. Rhod. Sci. Assoc. **6**, 1: 45, 61 (1906).—Bogdan, E. Afr. Agric. Journ. **15**: 122 (1950); Rev. List Kenya Grass.: 16 (1958).—Sturgeon, Rev. List Grass. S. Rhod. in Rhod. Agric. Journ. **51**: 222 (1954).—Chippendall in Meredith, Grasses & Pastures of S. Afr.: 59 (1955).—Bor, Grass. B.C.I. & P.: 546 (1960). —Napper, Grass. Tangan.: 14 (1965).—W. D. Clayton, in F.T.E.A. Gramineae: 41, fig. 15 (1970). TAB. **11** fig. A. Type from ? Europe.

A rather robust annual. Culms up to 90 cm. tall, 2–4-noded, solitary or tufted, somewhat stout, erect or geniculately ascending, glabrous, smooth or scabrous

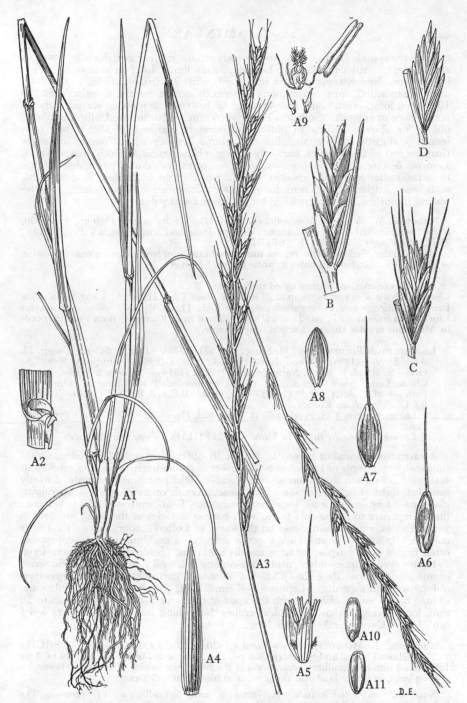

Tab. 11. A.—LOLIUM TEMULENTUM. A1, habit (× ⅔); A2, ligule (× 2); A3, inflorescence
(× 1); A4, superior glume (× 2); A5, floret (× 2); A6, floret showing rhachilla (× 2);
A7, lemma (× 2); A8, palea (× 2); A9, sexual organs with lodicules detached (× 6);
A10, A11, caryopsis showing embryo and hilum respectively (× 2), all from *Phipps*
2839. B.—LOLIUM TEMULENTUM var. ARVENSE, spikelet (× 2) *Monro* 492. C.—
LOLIUM MULTIFLORUM, spikelet (× 2) *Dod* 3562 D.—LOLIUM PERENNE, spikelet (× 2)
Chase 7676.

towards the spike. Leaf-sheaths prominently striate, rather tight, dorsally rounded, smooth or scaberulous, glabrous; Ligule c. 2 mm. long. Leaf-laminae 6–30(–40) × 0·3–1·3 cm., linear, tapering to a fine point, with 2 narrow auricles at the base, usually expanded, firm, scaberulous or smooth except for the margins. Spikes 10–30 cm. long, erect, rigid, pale-green; rhachis stout, somewhat waved, dorsally scaberulous or smooth. Spikelets 12–25 × 4–6 mm., 4–10-flowered, elliptic-oblong, oblong or obovate-oblong in outline, moderately compressed about their own length or slightly more apart. Superior glume 7–9-nerved, as long as or longer than the rest of the spikelet, narrowly oblong, obtuse, rigid, smooth or asperulous. Lemmas 6–8 mm. long, 5–9-nerved, muticous or awned, dorsally rounded, elliptic to ovate, somewhat rigid, smooth; awn (if present) up to 20 mm. long, straight, scabrous. Anthers c. 2·5 mm. long, linear. Caryopsis c. 7 mm. long, elliptic-oblong in outline, tightly enclosed by the lemma and palea.

Rhodesia. W: Victoria, *Monro* 492 (K; SRGH). C: Salisbury, 1460 m., 15.v.1930, *Brain* 1548 (K; SRGH). E: Melsetter, Umsapa-Jameson Farm, Musapa Valley, 1181 m., 21.ix.1960, *Phipps* 2839 (BM; K; SRGH).
Mainly in the Mediterranean region, introduced into most temperate regions. Growing as a weed in arable land and also in waste places. "Darnel"

Note: 2 varieties are distinguished in Europe:
Lemmas awned with stiff awns up to 20 mm. long (see TAB. **11**, fig. A.)   var. *temulentum*
Lemmas muticous or with very short awns (see TAB. **11**, fig. B)   -   -   var. *arvense*
Liljeblad in Svensk. Fl., ed. 3: 86 (1816). Type from Europe. Both varieties occur in Africa but are not always clearly distinguishable.

2. **Lolium multiflorum** Lam., Fl. Franç. **3**: 621 (1778).—Dur. & Schinz, Consp. Fl. Afr. **5**: 933 (1894).—Wood, Natal Pl. **5**, 4: t. 495 (1908).—Sturgeon, Rev. List Grass. S. Rhod. in Rhod. Agric. Journ. **51**: 222 (1954).—Jackson & Wiehe, Annot. Check List Nyasal. Grass.: 68 (1958).—Chippindall in Meredith, Grasses & Pastures of S. Afr.: 58, 59 (1955).—Bor, Grass. B.C.I. & P.: 545 (1960). TAB **11** fig. C. Type from Europe.
    Lolium scabrum J. S. Presl ex C. B. Presl, Rel. Haenk. **1**: 267 (1830). Type from Peru.
    Lolium italicum A. Braun in Flora, **17**: 243 (1834). Type from Europe.

A caespitose annual or biennial. Culms 30–80(100) cm. tall, 2–5-noded, sometimes solitary, simple or branched below, erect or geniculately ascending, smooth or scaberulous beneath the inflorescence. Leaf-sheaths prominently striate, dorsally rounded, tight at first, later somewhat loose, smooth or more rarely scaberulous, glabrous. Ligule 1–2 mm. long, membranous. Leaf-laminae 6–25 × 0·4–1 cm., linear, tapering to a fine soft point, with narrow auricles at the base, involute in young shoots, smooth and glossy on the lower and often scaberulous on the upper surface. Spikes 10–30 cm. long, erect or sometimes nodding, rather compressed, often tinged with purple; rhachis usually scabrous. Spikelets 7·5–25 mm. long, 5–15-flowered, elliptic-oblong to obovate-oblong in lateral view, glabrous. Superior glume much shorter than the spikelet, 4–7-nerved, narrowly oblong or lanceolate-oblong, obtuse to acute. Lemmas 5–8 mm. long, 5-nerved, awned, imbricate, oblong to lanceolate-oblong, with the apex obtuse and 2-dentate; awn up to 10 mm. long, straight, slender. Paleas scaberulous along the keels. Anthers 3–4·5 mm. long. Caryopsis 3–3·5 mm. long.

Rhodesia. E: Melsetter, Fairview Farm, c.1920 m.,15.xi.1950, *Crook* 284 (K; SRGH). Throughout Central and Southern Europe, NW. Africa and temperate regions of Asia. Introduced into almost all temperate areas of the rest of the world. Growing in lawns, on waste places, in arable land and along roads in higher rainfall areas.

Note: *L. multiflorum* is Italian Rye-grass; it easily hybridizes with *L. perenne*. The hybrid, the so-called " Short Rotation Rye-grass ", is regarded as a valuable pasture grass according to Bor (1960).

3. **Lolium perenne** L., Sp. Pl. **1**: 83 (1753).—Hack. in Engl. & Prantl, Nat. Pflanzenfam. **2**, 2: 77 (1887).—Dur. & Schinz, Consp. Fl. Afr. **5**: 933 (1894).—Monro in Proc. Rhod. Sci. Ass. **6**, 1: 45 (1906).—Chippindall in Meredith, Grasses & Pastures of S. Afr.: 59 (1955).—Bor, Grass. B.C.I. & P.: 545 (1960). TAB. **11**, fig. D. Type from Europe.
    Lolium brasilianum Nees, Agrost. Bras.: 43 (1829). Type from Brazil.

A caespitose perennial. Similar in habit to the preceding species. Leaf-laminae 3–20 cm. × 0·2–0·6 mm. Spikes 4–30 cm. long, stiff, slender to somewhat stout. Spikelets 7–20 mm. long, 4–14-flowered, elliptic to obovate-elliptic in lateral view. Superior glume shorter than the rest of the spikelet, 5–7-nerved, narrowly lanceolate to oblong-lanceolate, with the apex usually obtuse, smooth. Lemmas 5–7 mm. long, 5-nerved, muticous, imbricate, ovate-oblong or oblong, apex acute to sub-obtuse. Paleas with the keels scaberulous. Anthers 3–4 mm. long. Caryopsis c. 4·5 mm. long.

Rhodesia. C: Salisbury, 4·8 km. above Makabusi-Hunyani confluence on Makabusi R., 1400 m., 1.ix.1959, *Phipps* 2182 (BM; K; SRGH). E: Inyanga, London Farm, 1800 m., 7.x.1961, *Chase* 7676 (BM; SRGH).
Distributed throughout Europe, North Africa and the temperate areas of Asia, introduced into other temperate and subtropical parts of the globe. Growing as a weed in waste places, in arable land, roadsides, on sandy soil and on the edges of irrigated fields.

Note: This species, " Perennial Rye-grass ", is a widely cultivated pasture plant in Europe. It hybridizes with the preceding species. Whereas *L. perenne* grows as a perennial under normal conditions in Europe it may occasionally flower in the first year in Africa.

11. BRIZA L.

Briza L., Sp. Pl. **1**: 70 (1753); Gen. Pl., ed. 5: 32 (1754).

Spikelets pedicelled, solitary, several- to many-flowered, laterally compressed, densely imbricate. Florets all fertile, ☿, exserted from the glumes. Rhachilla disarticulating above the glumes and between the florets. Glumes 2, subequal, 3–9-nerved, persistent. Lemmas 7–9-nerved, broadly ovate to almost reniform from a deeply cordate base, chartaceous, with broad scarious margins all round, horizontally spreading. Paleas c. $\frac{2}{3}$–$\frac{4}{5}$ the length of the corresponding lemmas, membranous, 2-keeled, with the keels often narrowly winged. Lodicules 2, small, lanceolate-oblong. Stamens 3. Ovary glabrous. Caryopsis enclosed by the lemma and palea, dorsally compressed; hilum c. $\frac{1}{2}$ as long as the caryopsis, linear. Annual or perennial grasses. Ligules membranous. Inflorescence a loose contracted panicle.
A genus of c. 12 spp., distributed throughout Europe, temperate Asia and South America and North Africa.

Panicles with (1)3–12 spikelets, spikelets conspicuously large, 15–25 × 8–14 mm., 7–20-flowered - - - - - - - - - - - 1. *maxima*
Panicles with many spikelets, spikelets much smaller, 3–5 × 3–6 mm, 4–8-flowered
2. *minor*

1. **Briza maxima** L., Sp. Pl. **1**: 70 (1753).—Hack. in Engl. & Prantl, Nat. Pflanzenfam. **2**, 2: 72 (1887).—Dur. & Schinz, Consp. Fl. Afr. **5**: 899 (1895).—Stapf in Harv. & Sond., F.C. **7**: 708 (1900).—Sturgeon, Rev. List Grass. S. Rhod., in Rhod. Agric. Journ. **51**: 223 (1954).—Chippendall in Meredith, Grasses & Pastures of S. Afr.: 47, 49 (1955).—Bogdan, Rev. List Kenya Grass.: 15 (1958).—Bor, Grass. B.C.I. & P.: 527 (1960).—W. D. Clayton, in F.T.E.A., Gramineae: 53, fig. 19, 1–12 (1970). TAB. **12**, fig. A. Type from southern Europe.

A loosely tufted annual. Culms (10)20–70 cm. tall, 2–4-noded, simple, terete, slender, erect or ascending from a geniculate base, smooth. Leaf-sheaths often tinged with purple, loose, striate, dorsally rounded, glabrous, smooth. Ligule 2–5 mm. long. Leaf-laminae 5–25 × 0·25–0·75 cm., pale-green, tapering to an acute point, usually expanded, flaccid, glabrous, scaberulous along the edges. Panicles 4–10 cm. long, rather loose, with (1)3–12 spikelets, usually pendulous; axis very slender. Pedicels up to 2 cm. long, capillary, very flexible. Spikelets 15–25 × 8–14 mm., 7–20-flowered, pale-green to silver-green (often especially the glumes) purplish to red-brown, broadly ovate-cordate in outline, glabrous or often shortly pilose. Glumes 5–7 mm. long, often purplish, 7- (inconspicuously 9) nerved, broadly ovate from a cordate base, apex rounded to obtusely acuminate. Lemmas 6–7·5 mm. long, dorsally rounded, 7–9-nerved, broadly ovate, apex shortly acuminate, usually shortly pilose; the hairs tipped with glands. Paleas with the narrow wings shortly ciliate. Anthers 2–2·25 mm. long. Caryopsis c. 2·5 mm. long, light brown, broadly obovate to almost circular in outline.

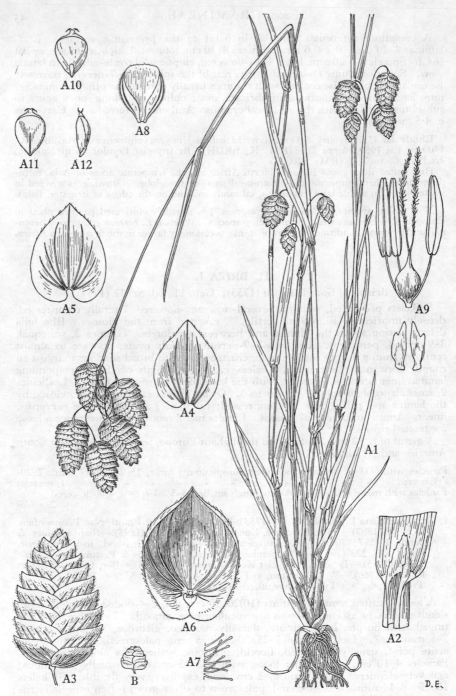

Tab. 12. A.—BRIZA MAXIMA. A1, habit (×⅔); A2, ligule (×2); A3, spikelet (×2); A4, inferior glume (×4); A5, superior glume (×4); A6, lemma (×4); A7, clavate hairs from lemma (×25); A8, palea (×4); A9, sexual organs with lodicules detached (×10), all from *Lotter* 14; A10, A11, A12, caryopsis in dorsal, ventral and lateral views (×4) from *Fourcade* 2355. B.—BRIZA MINOR, spikelet (×2) *Semsei* 1570.

Rhodesia. W: Matopos, xii.1925, *Eyles* 4077 (SRGH). E: Inyanga, 1710 m., 8.v.1963, *Rodel* s. n. (SRGH).

This grass, native to the Mediterranean region, has been introduced into parts of tropical Africa and S. Africa. In places it is naturalised in dry shady places and in grassland, also on roadsides and cultivated ground. It may also occur as an escape from gardens where it is grown as a most attractive ornamental grass (" Large Quaking Grass ").

2. **Briza minor** L., Sp. Pl. **1**: 70 (1753).—Hack. in Engl. & Prantl, Nat. Pflanzenfam. **2**, 2: 72 (1887).—Dur. & Schinz, Consp. Fl. Afr. **5**: 900 (1895).—Stapf in Harv. & Sond., F.C. **7**: 708 (1900).—Wood, Natal Pl. **5**, 4: t. 494 (1908).—Chippindall in Meredith, Grasses & Pastures of S. Afr.: 48, 49 (1955).—Verdcourt in Kew Bull. **10**: 599 (1956).—Bor, Grass. B.C.I. & P.: 528 (1960).—Napper, Grass. Tangan.: 13 (1965).—W. D. Clayton, in F.T.E.A., Gramineae: 53, fig. 19, 131 (1970). TAB. **12** fig. B. Type from Europe.

    Briza virens L., Sp. Pl., ed. 2, **1**: 103 (1762). Type from the Orient or Spain.

    Briza deltoidea Burm. f., Prodr. Fl. Cap.: 3 (1768). Type from Europe.

A loosely caespitose annual. Culms 10-60 cm. tall, erect or more rarely geniculately ascending, 2–4-noded, slender, terete, glabrous, smooth. Leaf-sheaths somewhat loose, rounded, smooth. Ligule 3–6 mm. long, obtuse. Leaf-laminae 3–14 × 0·3–0·9 cm., narrowly lanceolate, tapering to a fine point, usually expanded, asperulous on the upper surface and along the margins. Panicle 4–15(20) cm. long, obovate in outline, with many spikelets; branches spreading or obliquely ascending, scaberulous. Pedicels 5–15 mm. long, capillary. Spikelets 3–5 × 3–6 mm., 4–8-flowered, triangular-ovate to circular in outline, pendulous, sometimes erect, green or occasionally tinged with purple, glabrous, somewhat glossy. 2–3·5 mm. long, 3–5-nerved, apex obtuse or sometimes bluntly acuminate. Lemmas 2–3 mm. long, 7–9-nerved, deeply concave, apex broadly obtuse, dorsally rounded, indurated and glossy at the centre with broad whitish-hyaline margins. Paleas with the wings inconspicuously ciliolate. Anthers c. 0·6 mm. long. Caryopsis c. 2·25 mm. long, light brown, ventrally flat, dorsally convex.

Rhodesia. C: Marandellas, 26.ix.1931, *Rattray* 345 (BM; K; SRGH). E: Melsetter, gardens of Chimanimani Hotel, 1505 m., 20.ix.1960, *Phipps* 2833 (BM; K; LISC; SRGH).

Also in Tanzania. A native of the Mediterranean region, naturalised in temperate countries all over the world. Growing in cultivated land or waste ground, often as a garden escape. Sometimes grown for ornamental purposes but less frequently than the preceding species. In our area confined to the cooler and wetter regions.

POA 12. L.

Poa L., Sp. Pl. **1**: 67 (1753); Gen. Pl. ed. 5: 31 (1754).

Spikelets solitary, pedicelled, laterally compressed, muticous, 2–several-flowered; rhachilla disarticulating above the glumes and between the florets, terete, glabrous or sparsely hairy; florets all ☿ but for the usually reduced uppermost one (very rarely the 2 uppermost florets ♀ only). Glumes 2, slightly unequal, 1–3-nerved, thinly membranous, keeled, usually glabrous, persistent. Lemmas exceeding the glumes, (3)5–7-nerved, keeled, deeply concave, membranous, often with hyaline margins, apex acute to obtuse, usually pilose to a varying degree; callus short, obtuse, often with a woolly indumentum. Paleas shorter than or as long as the corresponding lemmas, 2-keeled, thinly membranous, usually ciliolate along the keels, with the apex usually emarginate. Lodicules 2, rather small, cuneate-2-lobate or lanceolate. Stamens 3. Ovary glabrous; styles distinct, rather short; stigmas plumose, laterally exserted. Caryopsis ovate, oblong or linear-oblong in outline, often grooved, usually free between the lemma and palea; embryo small; hilum basal, punctiform. Annuals or perennials of a wide range of habits. Ligule membranous or hyaline. Inflorescence a lax or contracted sometimes spike-like panicle.

A cosmopolitan genus of more than 200 spp. Mainly distributed throughout the temperate regions of the globe particularly in the northern hemisphere, but with a small number of spp. also represented in the tropics.

For critical comments on the taxonomy of the spp. which occur in the African mountains see Hedberg (p. 39 et seq.) whose concept has been accepted for this treatment with certain minor additions and modifications. In view of the small number of gatherings from our area a critical examination of the spp. is not yet possible.

Panicle contracted, narrow, almost linear in outline, with the branches erect and more or less appressed to the rhachis; spikelets densely clustered; loosely caespitose perennials
1. *leptoclada*

Panicle often with the branches drooping, spreading or ascending but never appressed to the rhachis; spikelets not densely clustered; perennials or annuals:

Lemmas with the apex acute in lateral view, acute to subacute when expanded; superior glume ovate-oblong to ovate-lanceolate, usually broadest ⅓ above the base; loosely caespitose perennials, growing in dense grassland or in open woodland, often in moist places - - - - - - - - - - - - 2. *schimperana*

Lemmas with the apex obtuse to obtuse-truncate in lateral view, obtuse to obtuse-retuse when expanded; superior glume ovate-elliptic to broadly elliptic, usually broadest at or somewhat above the middle; perennials or annuals:

Densely caespitose perennial; the lowest leaf-sheaths long, persistent and usually splitting into fibres; spikelets 3–4(5)-flowered; anthers c. 2 mm. long; montane grassland sp. - - - - - - - - - - - 3. *binata*

Loosely caespitose annuals (very rarely growing as perennials in higher altitudes); lowermost leaf-sheaths not splitting into fibres; spikelets 3–10-flowered; anthers 0·7–1·2(1·6) mm. long; weedy plant of waste places, cultivated ground or similar habitats - - - - - - - - - - - - 4. *annua*

1. **Poa leptoclada** Hochst. ex A. Rich., Tent. Fl. Abyss. **2**: 422 (1851).—Engl., Pflan-zenw. Ost-Afr. **A**: 124 (1895). op. cit. **C**: 115 (1895);—Eggeling, Annot. List Grass. Uganda: 39 (1947).—Sturgeon, Rev. List Grass. S. Rhod. in Rhod. Agric. Journ. **51**: 222 (1954).—Robyns & Tournay, Fl. Parc Nat. Alb. **3**: 187 (1955).—F. W. Andr., Fl. Pl. Sudan, **3**: 521 (1956).—Hedberg in Symb. Bot. Upsal. **15**: 39 (1957).—Bogdan, Rev. List Kenya Grass.: 15 (1958).—Jackson & Wiehe, Annot. Check List Nyasal. Grass: 54 (1958).—Harker & Napper, Ill. Guide Grass. Uganda: 50 (1960).—Napper, Grass. Tangan.: 13 (1965).—W. D. Clayton in F.T.E.A. Gramineae: 47 (1970). Type from Ethiopia. TAB. **13** fig. A.

Eragrostis puberula Steud., Syn. Pl. Glum. **1**: 268 (1854). Type from Ethiopia.

Poa pseudoschimperana Chiov. in Ann. Ist. Bot. Roma, **8**: 376 (1908). Type from Eritrea.

Poa schimperana var *longigluma* Chiov., tom. cit.: 377 (1908). Type from Eritrea.

Poa schimperana var. *micrantha* Chiov., loc. cit. Type from Eritrea.

Poa friesiorum Pilg. in Notizbl. Bot. Gart. Berl. **9**: 1127 (1927).—Robyns & Tournay, Fl. Parc Nat. Alb. **3**: 187 (1955). Type from Kenya.

Poa annua var. *hypsophila* Pilg., tom. cit.: 1126 (1927). Type from Kenya.

Poa schliebenii Pilg., op. cit. **12**: 384 (1935). Type from Kenya.

Poa oreades Peter, Fl. Deutsch. Ost-Afr. **1**, Anh.: 114 (1936). Type from Tanzania.

A loosely caespitose rather variable perennial. Culms 5–75 cm. tall, 2–5-noded, erect or geniculately ascending, branched below or solitary, very slender, wiry, terete, smooth, glabrous.   Leaf-sheaths slightly compressed, inconspicuously keeled, coarsely striate, loose, somewhat papery, pale-green to straw-coloured, the lowermost ones persistent and often splitting into irregular coarse fibres, smooth, glabrous, or scattered pilose.   Ligule (0·6)1–4(6) mm. long.   Leaf-laminae 2–12 × 0·05–0·25 cm., very narrowly linear, almost subulate, tapering to a fine soft point, usually folded or convolute, erect, somewhat curved, flaccid to somewhat firm, scaberulous along the nerves (more intensively so on the lower surface) or rarely smooth, usually quite glabrous, rarely scattered pilose.   Panicle (2·5)5–19 cm. long, linear or narrowly oblong in outline, almost spike-like, slender, erect, continuous or sometimes interrupted, straight or slightly curved, pale green or often tinged with purple; the branches almost always erect and appressed to the rhachis, scabrous to a varying degree.   Spikelets 3–5(6·8) mm. long, 2–5-flowered, elliptic to elliptic-oblong in outline.   Inferior glume 1·7–3·9 mm. long, 1-nerved, narrowly elliptic or elliptic-oblong, with the apex acute; the superior similar in shape but slightly longer, 3-nerved.   Lemma 2–3·2 mm. long (rarely longer), 5-nerved, elliptic, deeply cymbiform, with the apex acute, dorsally glabrous to pubescent, sometimes with a tuft of hair at the base.   Anthers 0·5–1 mm. long.

Rhodesia. E: Inyanga, 19.x.1946, *Rattray* 902 (SRGH). **Malawi. N**: near Lake Kaulima, Nyika Plateau, 2190 m., 14.ix.1962, *Tyrer* 958 (BM; COI; SRGH). **S**: Mlanje, near Lichenya Forestry Hut, 1770 m., 27.iii.1960, *Phipps* 2757 (BM; SRGH).

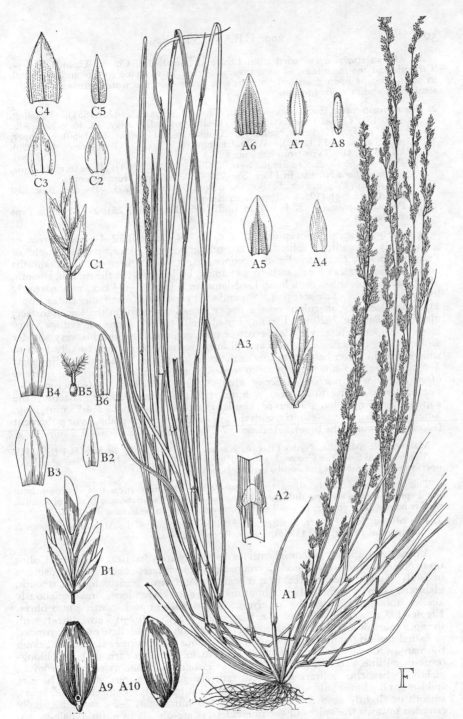

Tab. 13. A.—POA LEPTOCLADA. A1, habit (×½); A2, ligule (×1½); A3, spikelet (×6); A4, inferior glume (×6); A5, superior glume (×6); A6, lemma (×6); A7, palea (×6); A8, caryopsis with lodicules (×6); A9, caryopsis in dorsal view showing hilum (×18); A10, caryopsis in lateral view (×18), all from *Wiehe* 276. B.—POA ANNUA. B1, spikelet (×5); B2, inferior glume (×5); B3, superior glume (×5); B4, lemma (×5); B5, sexual organs (×6); B6, palea (×5), all from *Drummond* 5526. C.—POA SCHIMPERANA. C1, spikelet (×6); C2, inferior glume (×6); C3, superior glume (×6); C4, lemma (×6); C5, palea (×6), all from *Scaetta* 1565.

An African sp., also recorded from Ethiopia, Somaliland, Congo, Uganda, Kenya, Tanzania and the mountains of west tropical Africa. In montane forests and moorland, in semi-montane open grassland, in kloofs or along montane water-courses, always in semi-shade and in moist ground.

2. **Poa schimperana** Hochst. ex A. Rich., Tent. Fl. Abyss. **2**: 423 (1851).—Eggeling, Annot. List Grass. Uganda: 38 (1947).—Robyns & Tournay, Fl. Parc Nat. Alb. **3**: 190 (1955).—Harker & Napper, Ill. Guide Grass. Uganda: 50 (1960).—Napper, Grass. Tangan.: 13 (1965).—W. D. Clayton, in F.T.E.A., Gramineae: 48 (1970). TAB. **13** fig. C. Type from Ethiopia.

    Poa oligantha Hochst. ex Steud., Syn. Pl. Glum. **1**: 257 (1854). Type from Ethiopia.
    Poa viridiflora Hochst. in Flora, **38**: 232 (1855).—Type from Ethiopia.
    Poa perlaxa Pilg. in Notizbl. Bot. Gart. Berl. **9**: 1128 (1927).—Bogdan, Rev. List Kenya Grass.: 16 (1958). Type from Kenya.
    Poa muhavurensis C. E. Hubb. in Bull. Jard. Bot. Brux. **25**: 244 (1955). Type from Uganda.

A loosely caespitose perennial. Culms 15–70 cm. tall, 2–4-noded, erect or geniculately ascending, sometimes straggling, very slender, weak, simple or branched below, terete to slightly compressed, smooth, glabrous. Leaf-sheaths coarsely striate, rather loose, sometimes papery, keeled towards the mouth, smooth, glabrous. Ligule 0·5–6 mm. long. Leaf-laminae 3–18 × 0·1–0·4 cm., very narrowly linear, tapering to a fine soft point, expanded or convolute (often only in the upper $\frac{1}{2}$), rather flaccid, scaberulous on the upper, usually smooth on the lower surface, glabrous. Panicle 7–20 × 3–10 cm., broadly elliptic to oblong in outline, very loose and open, effuse, very lax, erect or nodding, with the rhachis very slender and curved; branches very delicate, filiform, solitary or paired, usually bare of spikelets in the lower half, spreading or deflexed, more rarely ascending. Spikelets 3–6 mm. long, 2–5-flowered, ovate to ovate-oblong in outline, pale green or more often tinged with purple. Inferior glume 1·75–4 mm. long, 1-(sometimes 3-) nerved, ovate-oblong to narrowly elliptic with the apex acute or subacute; the superior 2–4·5 mm. long, 3-nerved, similar in shape. Lemma 3·2–4·75 mm. long, 5-nerved, ovate to ovate-elliptic, with the apex acute, dorsally glabrous or pubescent (especially towards the base). Anthers 0·7–1 mm. long. Caryopsis c. 2 mm. long.

Malawi. N: Mzimba, Nyika Plateau, Kasaramba, 14.vi.1951, *Jackson* 504 (SRGH). Also recorded from the Congo, Uganda, Kenya and Tanzania. In montane grassland, open woodland and in forest scrub, usually in moist places.

This sp. is very closely allied to *P. leptoclada*. The only means of distinguishing the 2 spp. is by the striking difference of the panicles. *P. schimperana* is also of a much weaker habit. The ecology of the 2 spp., however, seems to be almost identical.

3. **Poa binata** Nees, Fl. Afr. Austr.: 378 (1841).—Chippindall in Meredith, Grasses & Pastures of S. Afr.: 53 (1955).—Type from S. Africa.

A densely caespitose perennial with leaves mostly crowded near the base. Culms up to 60 cm. tall, usually 2-noded, compressed below, terete above, slender, simple or rarely branched below, erect or rarely ascending from a geniculate base, smooth, glabrous. Leaf-sheaths rather tight at first, later somewhat loose, inconspicuously keeled, smooth, glabrous, the lowermost long persistent and splitting into fibres. Ligules 1–3 mm. long, obtuse. Leaf-laminae 2–15(18) × 0·2–0·5 cm., linear, with the apex more or less abruptly narrowed into a subobtuse cucullate point, expanded or folded around the midrib, smooth or scaberulous on the upper surface and along the margins, usually glabrous. Panicle 5–15 cm. long, ovate or triangular to oblong-ovate in outline, erect or nodding, open or rather lax; rhachis slightly grooved, glabrous; branches solitary or paired, rather slender, flexuous, divided, bare of spikelets in the lower half, spreading or deflexed at maturity. Pedicels rather short, smooth or slightly scaberulous. Spikelets 4–6 mm. long, 3–5-flowered, usually crowded towards the apices of the branchlets, ovate-oblong in outline, pale-green, sometimes slightly tinged with purple. Glumes unequal; the inferior 2–3 mm. long, 1-nerved but occasionally with 1 or 2 short additional lateral nerves, ovate-oblong or narrowly elliptic, apex subacute; the superior 3–4 mm. long, 3-nerved, ovate to ovate-elliptic, apex subobtuse, sometimes retuse. Lemmas 4–4·5 mm. long, 5-nerved, broadly elliptic or ovate-oblong, apex obtuse, often slightly retuse, hyaline, dorsally pubescent or glabrous. Paleas c. 3·5 mm. long, scabrous along the keels. Anthers c. 2 mm. long.

Rhodesia. E: Inyanga, Mt. Inyangani, 2130–2440 m., 23.ix.1957, *Phipps* 736 (K; SRGH).

Also distributed over the mountainous areas of the eastern Cape Province, eastern Transvaal, Natal and Lesotho. On slopes of mountains, on hills, in crevices of rocks and in open mountain grassland.

4. **Poa annua** L., Sp. Pl. **1**: 68 (1753).—Stapf in Harv. & Sond., F. C. **7**: 712, 715 (1900). —Eggeling, Annot. List Grass. Uganda: 39 (1947).—Sturgeon, Rev. List Grass. S. Rhod. in Rhod. Agric. Journ. **51**: 222 (1954).—Robyns & Tournay, Fl. Parc Nat. Alb. **3**: 188 (1955).—Chippindall in Meredith, Grasses & Pastures of S. Afr.: 52 (1955).—Hedberg in Symb. Bot. Upsal. **15**: 40 (1957).—Bogdan, Rev. List Kenya Grass.: 16 (1958).—Harker & Napper, Ill. Guide Grass. Uganda: 50 (1960).—Bor, Grass. B.C.I. & P.: 555 (1960).—Napper, Grass. Tangan.: 13 (1965).—W. D. Clayton in F.T.E.A., Gramineae: 49 (1970). TAB. **13** fig. B. Type from Europe.
Poa bipollicaris Hochst. in Flora, **38**: 321 (1855). Type from Ethiopia.

A caespitose delicate annual (very rarely perennial). Culms 2·5–30 cm. tall, 2–4-noded, erect, spreading or decumbent, sometimes rooting at the lower nodes, rather slender, terete, glabrous, smooth. Leaf-sheaths usually loose, somewhat compressed, keeled, smooth, usually glabrous. Ligule 2–5 mm. long. Leaf-laminae 1·5–10(15) × 0·1–0·5 cm., linear or linear-oblong, with the apex abruptly acute or often cucullate, expanded or folded, flaccid, scaberulous along the margins otherwise smooth, glabrous. Panicle 2–8(12) cm. long, open or more rarely contracted, ovate in outline, somewhat stiff; branches usually 2 (rarely 3–5)-nate or solitary, spreading at first, later usually slightly deflexed, smooth; pedicels 0·3–5 mm. long, smooth. Spikelets 3–8(10) mm. long, 3–7(10)-flowered, ovate to oblong in outline, usually crowded, green or often tinged with purple. Inferior glume 1·5–3 mm. long, 1-nerved, ovate or lanceolate; the superior 2–4·25 mm. long, 3-nerved, elliptic or oblong. Lemmas 2·3–4·25 mm. long, broadly elliptic or ovate-elliptic with white hyaline margins, with the apex obtuse. Anthers 0·7–1·2(1·6) mm. long. Caryopsis 1–2 mm. long.

Zambia. N: Abercorn, 1620 m., 27.vi.1963, *Vesey-FitzGerald* 4154 (K; SRGH). S: Choma, 1310 m., 28.v.1955, *Robinson* 1262 (K; SRGH). **Rhodesia.** W: Bulawayo, Municipal Park, 1370 m., 30.iv.1958, *Drummond* 5526 (BM; LISC; SRGH). E: Melsetter, around Chimanimani Mt. hut, 1650 m., 26.x.1959, *Goodier & Phipps* 298 (BM; SRGH).

A cosmopolitan sp., certainly introduced. In waste places, disturbed areas, along roads and paths, in forest clearings, as a weed of cultivation, rarely in montane grasslands or open woodlands. Usually confined to high rainfall areas except where artificially watered.

13. PSEUDOBROMUS K. Schum.*

Pseudobromus K. Schum. in Engl., Pflanzenw. Ost-Afr. **C**: 108 (1895).

Spikelets solitary, pedicelled, slightly compressed laterally, 1–4-flowered, awned; rhachilla disarticulating above the glumes and (if several-flowered) between the florets, always produced and terminated by a reduced awned lemma (often apparently only a bristle-like awn); florets 1 or 2(4), usually all ♀ or (perhaps?) the uppermost ♂. Glumes 2, persistent, subequal, 1–3-nerved, membranous. Lemmas exceeding the glumes, 3–sub-5-nerved (very rarely fully 5-nerved), firmly membranous, dorsally rounded, awned from the apex or sometimes from between 2 short acute apical lobes; awn straight, slender, terete. Paleas c. as long as the body of the lemmas, 2-nerved, 2-keeled, membranous, rather narrow, with the keels close together, with the apex entire or nearly so. Lodicules 2, oblanceolate-cuneate, with the apex usually 2-lobed. Stamens 3. Ovary oblong, with the apex distinctly pilose; styles short, distinct; stigmas plumose, laterally exserted. Caryopsis oblong in outline, ventrally longitudinally grooved, free between the lemma and the palea, with the apex distinctly pilose. Perennial forest grasses with usually flaccid leaves. Leaf-laminae usually inconspicuously and

* The genus *Pseudobromus* is not easily distinguishable from *Festuca*. The fact, however, that the lemmas are predominantly 3-nerved and the leaves cross-veined to a varying degree makes it advisable to maintain it as a separate genus. Moreover 1-flowered spikelets are virtually unknown in the genus *Festuca*. According to Metcalfe (p. 414) the affinities of *Pseudobromus* lie with *Bromus* rather than *Festuca*.

irregularly tessellate. Ligule a membrane. Inflorescence a large, open panicle, erect or nodding.

An African genus (including Madagascar) of 3 or 4 spp.

Spikelets 1-flowered with the rhachilla produced beyond the floret into a fine bristle (sometimes terminating in a reduced lemma, very rarely with a complete second floret); inferior glume 2·5–4·5(5·25) mm. long, the superior (3)4·5–6·75 mm. long
1. *silvaticus*
Spikelets 3–4-flowered with the 4th floret sometimes reduced to a lemma; inferior glume 4·5–8 mm. long, the superior 6–11 mm. long - - - - 2. *engleri*

1. **Pseudobromus silvaticus** K. Schum. in Engl., Pflanzenw. Ost-Afr. **C**: 108 (July 1895).—Eggeling, Annot. Check List Grass. Uganda: 39 (1947).—C. E. Hubbard in Mem. N. Y. Bot. Gard. **9**, **1**: 103 (1954).—Robyns & Tournay, Fl. Parc. Nat. Alb. **3**: 197 (1955).—F. W. Andr., Fl. Pl. Sudan, **3**: 521 (1956).—Jackson & Wiehe, Annot. Check List Nyasal. Grass.: 55 (1958).—Bogdan, Rev. List Kenya Grass.: 15 (1958).—Harker & Napper, Ill. Guide Grass. Uganda: 51 (1960).—Napper, Grass. Tangan.: 15 (1965).—W. D. Clayton in F.T.E.A., Gramineae: 54 (1970). Type from Tanzania.
Brachyelytrum africanum Hack. in Bull. Herb. Boiss. **3**: 382 (August 1895). Type from S. Africa.
Pseudobromus africanus (Hack.) Stapf in Dyer, F. C. **7**: 763 (1900).—Stent in Bothalia, **1**: 278 (1924).—Phillips, S. Afr. Grass.: 174, pl. 55 (1931).—de Winter in Bothalia, **6**: 140 (1951).—Chippindall in Meredith, Grasses & Pastures of S. Afr.: 61 (1955). Type as above.

A caespitose perennial, with a short knotty rhizome. Culms up to 125 cm. tall, 2–7-noded, branched from the base, erect or sharply ascending, slender, wiry, faintly striate, usually glossy, glabrous. Leaf-sheaths slightly longer than the internodes, tight at first, later often slipping off the culm, scaberulous or smooth. Ligule 3–8 mm. long, obtuse to acuminate, glabrous. Leaf-laminae 5–45 × 0·3–1·5 cm., narrowly lanceolate to linear, base auricled, tapering to a setaceous point, usually expanded, rarely folded around the middle, somewhat rigid, dark green, scaberulous on the nerves especially on the upper surface, scabrous along the margins. Panicle up to 45 cm., long, terminal, lax, open, slightly nodding or erect; rhachis rather slender, terete or sometimes flattened, smooth or often scaberulous towards the apex; branches up to 20 cm. long, fascicled, rarely solitary, terete, flexuous, sparsely branched. Spikelets 7–10 mm. long, 1-flowered, with the rhachilla produced beyond the floret into a fine bristle (sometimes terminating in a reduced lemma), narrowly elliptic or narrowly ovate in outline, dark green, slightly laterally compressed to terete; extension of the rhachilla slightly longer than ½ the length of the spikelet. Glumes with the apex acute or rarely subobtuse, margins hyaline; the inferior 2·5–4·5(5·25) mm. long, 1-nerved (very rarely sub-3-nerved), lanceolate triangular; the superior (3)4·5–6·75 mm. long, 3-nerved (the lateral nerves sometimes evanescent), broadly lanceolate-triangular. Lemma 6–9 mm. long (excluding awn), 3-nerved, narrowly lanceolate; awn (8)14–20 mm. long. Palea c. as long or slightly longer than the lemma. Callus rounded, pilose, gibbous in lateral view. Anthers 2·5–4·75(6) mm. long. Caryopsis c. 6 mm. long, reniform in cross-section; embryo ⅙–⅕ the length of the caryopsis; hilum linear, stretching the whole length of the caryopsis.

Zambia. E: Lundazi, Nyika Plateau, Chowo Forest, ix.1968, *Williamson* 1011 (BM; K). **Malawi.** N: Nyika Plateau, 3·2 km. SW. of Rest Hut, 21.x.1958, *Robson* 199 (BM; K; LISC). C: Dedza, Cirobwe Forest, 22.v.1961, *Chapman* 1317 (BM; K; LISC; NYAS; SRGH). S: Mlanje Mt., Lichenya Plateau, 1850 m., 6.vi.1962, *Robinson* 5278 (BM; K; LISC; SRGH). **Mozambique.** Z: Gúruè on the top of the Serra do Gúruè, 20.ix.1944, *Mendonça* 2157 (LISC). MS: Manica, Tsetsera, 2130 m., 30.i.1960, *Mackenzie* 3080 (BM; SRGH).

Also recorded from the Sudan, Uganda, Kenya, Tanzania, the north-eastern Cape Province, northern Transvaal, Natal and Zululand. Growing in evergreen forests in places where the shade is not too heavy, also on banks in forest streams.

2. **Pseudobromus engleri** (Pilg.) W. D. Clayton in Kew Bull. **23**: 293 (1969); in F.T.E.A., Gramineae: 54, fig. 20 (1970). TAB. **14.** Type from Tanzania.
Festuca engleri Pilg. in Engl., Bot. Jahrb. **40**: 85 (1907). Type as above.

Tab. 14. PSEUDOBROMUS ENGLERI. 1, habit (× ½) *Jackson* 863; 2, ligule (× ⅔); 3, inflorescence-branchlet bearing a spikelet, and a pair of persistent glumes (× 1⅓); 4, inferior glume (× 2⅔); 5, superior glume (× 2⅔); 6, spikelet, with glumes removed (× 1⅓); 7, detail from base of floret (× 4); 8, lemma (× 2⅔); 9, palea (× 2⅔); 10, sexual organs with lodicules detached (× 2⅔), 2–10 from *Jackson* 519; 11, caryopsis, dorsal and ventral views (× 2⅔) *Jackson* 863.

Pseudobromus brassii C. E. Hubb. in Kew Bull. **4**: 341 (1949); in Mem. N.Y. Bot. Gard. **9**, 1: 102 (1954).—Jackson & Wiehe, Annot. Check List Nyasal. Grass.: 55 (1958). Type: Malawi, Nyika Plateau, *Brass* 17282 (K, holotype; SRGH, isotype).
Festuca gigantea var. *africana* Robyns & Tournay in Bull. Jard. Bot. Brux. **25**: 245 (1955); in Fl. Parc Nat. Alb. **3**: 196 (1955). Type from Kenya.
Festuca gigantea sensu Napper, Grass. Tangan.: 14 (1965).

A caespitose perennial, in habit quite similar to the preceding sp. Culms up to 180 cm. tall. Spikelets 8–12 mm. long, 2–4-flowered (if only 2-flowered then with the second floret fully developed, ♂ (?) or usually ♀). Inferior glume 4·5–8 mm. long, 1 (very rarely sub-3)-nerved; the superior 6–11 mm. long, 3-nerved. Lemma 7–14 mm. long (excluding awn), 3–sub-5-nerved; awn 10–16 mm. long. Anthers 2·5–5 mm. long.

Rhodesia. E: Inyanga, 1460 m., 24.ii.1961, *Phipps* 2859 (K; SRGH). **Malawi.** N: Nyika, 8.vii.1962, *Lawton* 911 (SRGH). S: Mt. Mlanje, Chalonwe, 1830 m., 29.iii.1960, *Phipps* 2779 (SRGH).
Also recorded from Tanzania. In openings of montane forests, often along forest streams.

14. FESTUCA L.*

Festuca L., Sp. Pl. **1**: 73 (1753); Gen. Pl. ed. 5: 33 (1754).

Spikelets several–many-flowered, laterally compressed, pedicelled, solitary; rhachilla slender, slightly elongated, disarticulating above the glumes and between the florets. Florets clearly exserted from the glumes, ♀ but the uppermost one usually reduced; callus small, annular, usually glabrous. Glumes subequal, sometimes the inferior smaller, herbaceous to subcoriaceous, keeled; the inferior 1–3-nerved; the superior 3-nerved. Lemmas lanceolate, (3-)5-(7)nerved, herbaceous to subcoriaceous, dorsally rounded or rarely somewhat keeled towards the apex, usually awned or with an awn-point from the apex, rarely awnless or awned dorsally from below the apex. Paleas as long as or slightly shorter than the corresponding lemmas, 2-keeled, usually with the apex 2-dentate, thinly membranous or hyaline, with the keels scaberulous or ciliolate. Lodicules 2, entire or asymmetrically 2-lobed. Stamens 3. Ovary glabrous or sometimes with the apex pubescent; styles 2, laterally exserted. Caryopsis usually elongate, usually free but tightly enclosed by the lemma and the palea, dorsally compressed and grooved; embryo small, not reaching ⅓ the length of the caryopsis; hilum basal. Caespitose perennials, often rhizomatous. Leaf-laminae folded or convolute, often setaceous, rarely expanded. Ligule scarious or sometimes membranous, usually entire. Inflorescence an open or contracted panicle.

A genus of almost 100 spp., distributed throughout the temperate and subtropical regions of both hemispheres, represented by a few spp. in mountain areas of the tropics.

Panicle condensed, spiciform or almost linear, with the branches short and appressed to the axis; spikelets 6·5–10(12) mm. long, not gaping - - - 1. *abyssinica*
Panicle loose and open or somewhat contracted, often slightly drooping, with the branches long and flexible; spikelets 10–20 mm. long, gaping when mature:
  Lemmas acute but not awned, sometimes very shortly (less than 1 mm.) mucronulate, the lowermost 6·5–10 mm. long; inferior glume usually 3-nerved, lanceolate. Spikelets 3–6-flowered - - - - - - - - 2. *costata*
  Lemmas always awned from an entire or shortly 2-lobed apex; the lowermost 5–7·5 mm. long; inferior glume 1-nerved, narrowly lanceolate to subulate. Spikelets 4–7(9)-flowered - - - - - - - - 3. *caprina*

* *Festuca arundinacea* Schreb., Spic. Fl. Lips.: 57 (1771) is a temperate sp. which has been introduced as a pasture grass into various parts of Africa. It can be distinguished from the other spp. of *Festuca* of the Flora Zambesiaca area by the minutely hairy falcate auricles at the base of the leaf-laminae. According to Jackson & Wiehe (p. 67) this sp. is under trial in montane grassland areas of Malawi. It has also been recorded from Rhodesia.

1. **Festuca abyssinica** A. Rich., Tent. Fl. Abyss. **2**: 433 (1851).—Engl., Pflanzenw. Ost-Afr. **A**: 126 (1895); op. cit. **C**: 116 (1895);—Eggeling, Annot. List Grass. Uganda: 21 (1947).—Robyns & Tournay, Fl. Parc Nat. Alb. **3**: 192 (1955).— Bogdan, Rev. List Kenya Grass.: 15 (1958).—Harker & Napper, Ill. Guide Grass. Uganda: 34 (1960).—Napper, Grass. Tangan.: 14 (1965).—W. D. Clayton in F.T.E.A., Gramineae: 60, fig. 21 (1970). TAB. **15**. Type from Ethiopia.

Festuca schimperana A. Rich., Tent. Fl. Abyss. **2**: 433 (1851).—Eggeling, Annot. List Grass. Uganda: 21 (1947).—Jackson & Wiehe, Annot. Check List Nyasal. Grass.: 42 (1958).—Type from Ethiopia.

Festuca restituta Steud., Syn. Pl. Glum. **1**: 314 (1854) *nom. superfl.* Based on *Festuca schimperana*; however the description refers to *Festuca simensis*.

Festuca rigidula Steud., loc. cit.—Eggeling, Annot. List Grass. Uganda: 21 (1947). —F. W. Andr., Fl. Pl. Sudan, **3**: 461 (1956). Type from Ethiopia.

Festuca abyssinica var. *acuta* Rendle, Cat. Afr. Pl. Welw. **2**: 255 (1899). Type from Angola.

Festuca gelida Chiov. in Ann. Bot. Roma, **6**: 147 (1907).—Type from Congo Republic.

Festuca abyssinica subsp. *acamptophylla* St. Yves in Notizbl. Bot. Gart. Berl. **9**: 1132 (1927). Type from Tanzania.

Festuca abyssinica var. *intermedia* St. Yves, loc. cit. Type from Kenya.

Festuca abyssinica var. *keniana* St. Yves, loc. cit. Type from Kenya.

Festuca abyssinica var. *schimperana* St. Yves, loc. cit. Type as for *Festuca schimperana*.

Koeleria afromontana Jacques-Félix, Gram. Afr. Trop.: 186 (1962) *nom. nud.*

A loosely to densely caespitose perennial, rather variable. Culms (15)25–60 (80) cm. tall, 2–5-noded, geniculately ascending, sometimes straggling, rarely erect, terete, somewhat weak, glabrous, smooth. Leaf-sheaths striate, open, tight when young but soon slipping off the culm, usually glabrous; the older ones splitting into irregular brown fibres. Ligule 0·5–1 mm. long. Leaf-laminae 4–27 cm. long, up to 3 mm. wide, soft, usually involute, filiform or even acicular. Panicle 7–25 cm. long, rather dense, narrowly oblong to linear in outline, often spike-like; branches short, usually closely appressed to the rhachis, scaberulous. Spikelets shortly pedicelled, 6·5–10(12·5) mm. long, 2–6-flowered, bright to olive green, sometimes tinged with purple, oblong-ovate in outline, not gaping. Glumes membranous, with thinner margins, embracing the spikelet fairly tightly to $\frac{2}{3}$ to $\frac{3}{4}$ of its length, acute; the inferior 5–8·25 cm. long, 1–3-nerved, lanceolate, slightly asymmetric; the superior 6–10 mm. long, 3 (sometimes 5-)nerved, lanceolate to ovate-lanceolate. Lemmas 6–9·5 mm. long, narrowly elliptic to ovate-oblong, acute, subacute but usually tapering to an awn point up to 5 mm. long, dorsally scaberulous. Paleas with the keels scaberulous. Anthers 2–2·5 mm. long. Ovary glabrous or with the apex pilose.

Zambia. N: Luwingu, Chishinga Ranch, 1410 m., 13.ix.1961, *Astle* 908 (BM; SRGH). **Rhodesia.** E: Inyanga, on slope of Inyangani up to summit ridge, 30.iv.1965, *West* 6432 (K., SRGH). **Malawi.** N: Rumpi, Chelinda, 26.v.1967, *Salubeni* 733 (BM; SRGH). S: Mlanje Mt. near Lichenya, 1740 m., 27.iii.1960, *Phipps* 2754 (K; PRE; SRGH). **Mozambique.** MS: Manica, Tsetsera, 2220 m., 3.iii.1954, *Wild* 4468 (K; PRE; SRGH).

Known from the Cameroun Republic, Ethiopia, Uganda, Kenya and Tanzania. Growing in mountain grassland, in moist and often peaty soils.

Festuca abyssinica is an extremely polymorphic sp. of which quite a number of infra-specific taxa have been described, but all of them intergrade to such a degree that it is virtually impossible to key them out. It has often been confused in the herbarium as well as in the field with *Koeleria capensis* (see. p. 69) but this plant can easily be distinguished from *Festuca abyssinica* by its pubescent to almost tomentose rhachis of the panicle.

2. **Festuca costata** Nees, Fl. Afr. Austr.: 447 (1841).—Engl., Pflanzenw. Ost-Afr. **A**: 128, 132 (1895); op. cit. **C**: 116 (1895);—Stapf in Harv. & Sond., F.C. **7**: 721 (1900).—Sturgeon, Rev. List Grass. S. Rhod. in Rhod. Agric. Journ. **51**: 221 (1954).—Chippindall in Meredith, Grasses & Pastures of S. Afr.: 56 (1955).—Bogdan Rev. List Kenya Grass.: 15 (1958).—Napper, Grass. Tangan.: 14 (1965).— W. D. Clayton in F.T.E.A., Gramineae: 59 (1970).—Type from S. Africa.

Festuca milanjiana Rendle in Trans. Linn. Soc., Ser. 2, **4**: 59 (1894).—Engl., Pflanzenw. Ost-Afr. **A**: 132 (1895); **C**: 116 (1895); op. cit.—Jackson & Wiehe, Annot. Check List Nyasal. Grass.: 42 (1958). Type. Malawi, Mt. Mlanje. *Whyte* (BM; holotype; K, isotype).

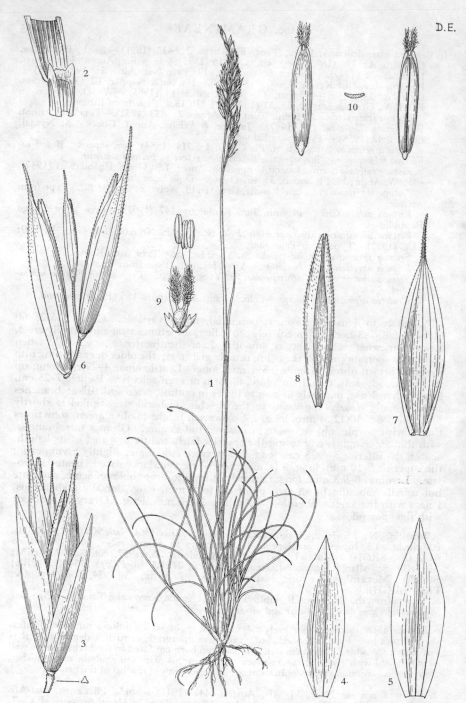

Tab. 15. FESTUCA ABYSSINICA. 1, habit (×½); 2, ligule (×5); 3, spikelet (×6); 4, inferior glume (×6); 5, superior glume (×6); 6, spikelet with glumes removed (×6); 7, lemma (×6); 8, palea (×6); 9, sexual organs with lodicules (×6); 10, caryopsis, dorsal and ventral views, and transverse section (×6), all from *Hedberg* 414. From F.T.E.A.

Densely tufted perennial. Culms 50–120(150) cm. tall, 1–3-noded, stiffly erect, terete, growing from a slightly thickened base, glabrous, smooth. Leaf-sheaths finely striate, open, fairly tight; the older ones remaining, splitting into regular thin brown fibres, which conspicuously surround the culm-bases. Ligule up to 4 mm. long. Leaf-lamina 10–60 cm. long, 2–4 mm. wide, usually involute, very rarely expanded, tapering to a setaceous point, erect or obliquely ascending, glabrous, smooth but scaberulous along the edges. Panicle up to 25 cm. long, loose and open, rarely somewhat contracted, erect or slightly drooping; branches usually 2-nate, rather slender, flexuous, branched above the middle, the lower ones fairly remote. Pedicels of unequal length, filiform. Spikelets 10–20 mm. long, 3–6-flowered, green but often tinged with purple, gaping, often drooping. Glumes 3-nerved, unequal, lanceolate, acute; the inferior 3·5–6·6 mm. long; the superior 4–8 mm. long. Lemmas 6–10 mm. long, chartaceous, lanceolate, dorsally scaberulous, with the apex subobtuse, acute or very shortly mucronulate from below two minute apical lobes. Paleas with the keels scabrous. Anthers c. 3·5 mm. long. Ovary glabrous or pubescent at the apex.

Rhodesia. E: Inyanga, c. 6·4 km. from Troutbeck Hotel, viii.1959, *Williams* 159 (BM; SRGH). **Malawi.** S: Mt. Mlanje, Lichenya Plateau, 2000 m., 11.vi.1962, *Robinson* 5333 (K; SRGH). **Mozambique.** MS: Manica, Serra Zuira, Tsetsera plateau, at 1·5 km. from the dairy road to Vila Pery, 2100 m., 4.xi.1965, *Torre & Pereira* 12633 (LISC).

Distributed throughout eastern Africa from Kenya southwards and extending into Natal, Lesotho, Swaziland and the eastern Transvaal. In montane grassland between 2400 and 3000 m., sometimes also found in forest clearings.

3. **Festuca caprina** Nees, Fl. Afr. Austr.: 443 (1841).—Stapf in Harv. & Sond., F. C. **7**: 719 (1900).—Sturgeon, Rev. List Grass. S. Rhod. in Rhod. Agric. Journ. **51**: 221 (1954).—Chippindall in Meredith, Grasses & Pastures of S. Afr.: 55, 56 (1955). —Jackson & Wiehe, Annot. Check List Nyasal. Grass.: 42 (1958).—W. D. Clayton in F.T.E.A., Gramineae: 62 (1970). TAB. **16**. Type from S. Africa.

Festuca nubigena subsp. *caprina* (Nees) St. Yves in Rev. Bret. Bot. **2**: 79 (1927). Type as above.

A very densely caespitose perennial. Culms 30–100 cm. tall, 1–2(3)-noded, erect, slender, wiry, smooth, glabrous. Leaf-sheaths fairly tight, open towards the mouth, smooth; the lower persistent, splitting into brown fibres which form more or less dense cushions around the culm-bases. Ligule very short, usually less than 1 mm. long. Leaf-laminae 3–25 × 0·05–0·2 cm., usually involute and filiform, rarely expanded, smooth, or inconspicuously scaberulous on the upper surface. Panicle 5–20 cm. long, open, rarely slightly contracted, narrowly ovate in outline, erect or somewhat drooping; branches slender but less flexuous than in the preceding sp., single or paired, simple or branched near the base. Pedicels unequal, short. Spikelets 10–15 mm. long, 4–7(9-)-flowered, oblong, gaping, bright green or tinged with dark purple, usually erect, rarely drooping. Glumes unequal, chartaceous, acute; the inferior 3–5·5 mm. long, 1-nerved, narrowly lanceolate to subulate; the superior 3·5–6·5 mm. long, 3-nerved, lanceolate. Lemmas 5–7·5 mm. long, dorsally scaberulous, sometimes thinly pubescent, lanceolate, with the apex tapering into a fine awn 1–3 mm. in length. Palea with the keels scabrous. Anthers 2·5–3 mm. long. Ovary shortly pilose at the apex.

Zambia. N: Zombe, Abercorn, Muyumvya dambo, 1650 m., 26.i.1961, *Vesey-FitzGerald* 2940 (SRGH). **Rhodesia.** E: Inyanga, near Troutbeck, 2100 m., 21.x.1946, *Rattray* 924 (K; SRGH). **Malawi.** N: Nyika Plateau, in valley-bottom near stream, 22.ii.1949, *Hawksworth* in GHS 22682 (K; SRGH).

Widely distributed in S. Africa (eastern Cape Prov., Lesotho, Natal and eastern Transvaal) extending through Rhodesia and Malawi into Tanzania. In montane grassland above 1600 m., often in valley bottoms near streams.

VIII. BROMEAE Dumort.

Bromeae Dumort., Obs. Gram. Belg.: 82, 115 (1823).

Inflorescence a panicle, open or contracted. Spikelets all alike, laterally compressed to subterete, few–many-flowered; florets ☿ but the upper ones often reduced; rhachilla disarticulating above the glumes and below each floret. Glumes

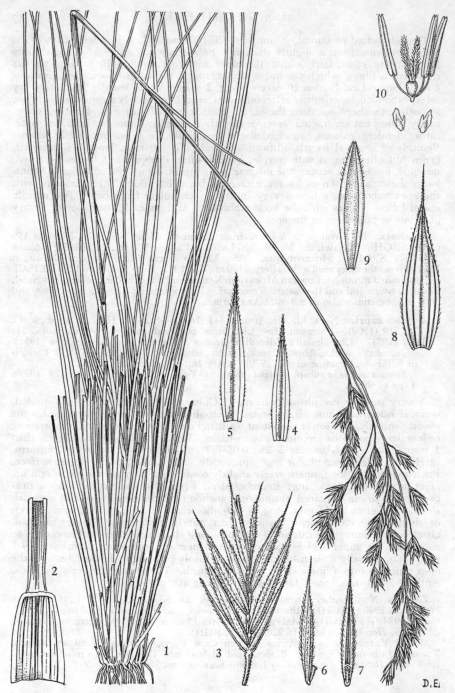

Tab. 16. FESTUCA CAPRINA. 1, habit (×⅔) *Rattray* 924 and *Crook* 280; 2, ligule (×4) *Crook* 280; 3, spikelet (×4); 4, inferior glume (×6); 5, superior glume (×6); 6, 7, floret, lateral and ventral views respectively (×6); 8, lemma (×6); 9, palea (×6); 10, sexual organs with lodicules detached (×6), 3–10 from *Rattray* 924.

2, persistent, unequal, shorter than the lowest floret, herbaceous to chartaceous, often with scarious margins; the inferior 1–3-nerved; the superior 3–7-nerved. Lemmas 5–9-nerved, dorsally rounded, usually 2-dentate, awned from below the apices; awn straight or recurved, sometimes twisted. Paleas usually shorter than the corresponding lemmas, 2-nerved, 2-keeled, hyaline, with the keels ciliate or scabrous. Lodicules 2. Stamens 3 or 2. Ovary with an apical pilose appendage; styles 2, rather short, arising laterally from the base of the appendage. Caryopsis elongate, usually convex-concave in cross-section; embryo small; hilum linear, almost extending the entire length of the caryopsis; starch grains simple, rounded. Chromosomes large, basic number 7.

A tribe of 2 genera, distributed mainly throughout the temperate regions of the northern hemisphere, but also present in tropical highlands.

15. BROMUS L.

Bromus L., Sp. Pl. **1**: 76 (1753); Gen. Pl., ed. 5: 33 (1754).

Spikelets pedicelled, solitary, 2–many-flowered, laterally compressed to a varying degree, usually compact; rhachilla disarticulating above the glumes and between the florets; florets all ♀ but the uppermost one usually reduced. Glumes 2, unequal, or very rarely equal, similar, persistent, keeled, membranous, with the apex subacute; the inferior 1–7-nerved; the superior 3–9-nerved. Lemmas 5–9-(–13)nerved, keeled, membranous, lanceolate to lanceolate-elliptic, with the apex 2-dentate or 2-lobed, awned or mucronate from the sinus, usually scabrous along the keel and the nerves. Paleas c. as long as or slightly shorter than the corresponding lemmas, thinly membranous, 2-nerved, 2-keeled, linear or lanceolate, with the apex entire or 2-fid, ciliate or ciliolate or pectinate along the keels. Lodicules 2, entire or lobed. Stamens 3. Ovary obovate in outline, crowned by an apical pilose or villous appendage. Styles 2, distinct, inserted laterally below the appendage, rather short; stigmas plumose, laterally exserted (permanently enclosed in cleistogamic spp.). Caryopsis linear-oblong in outline, usually concave-convex in cross-section, adherent to the palea; embryo small, basal; hilum linear, stretching almost the entire length of the caryopsis. Annuals or perennials. Ligule membranous. Inflorescence a lax or contracted panicle (rarely reduced to a raceme)

A large genus of c. 100 spp., distributed throughout the temperate areas and the montane regions of the tropics.

Lemmas muticous or with a short awn not exceeding 3 mm., 9–13-nerved, sharply keeled. Inferior glume 5–7-nerved; the superior 7–9-nerved; spikelet strongly laterally compressed, appearing almost flat   -   -   -   -   -   1. *unioloides*
Lemmas with an awn up to 18 mm. long, 3–7-nerved, not sharply keeled. Inferior glume 1-nerved; the superior 3-nerved. Spikelet less compressed   -   -   2. *leptocladus*

1. **Bromus unioloides** Kunth, Nov. Gen. & Sp. Pl. **1**: 151 (1816).—Hack. in Engl. & Prantl, Nat. Pflanzenfam. **2**, 2: 76 (1887).—Stapf in Harv. & Sond., F. C. **7**: 727, 734 (1900).—Wood, Natal Pl. t. **5**, 3: pl. 461 (1906).—Sturgeon in Rhod. Agric. Journ. **51**: 225 (1954).—C. E. Hubb. in Agron. Lusit. **18**: 7 (1956).—Bogdan, Rev. List Kenya Grass.: 14 (1958).—Bor, Grass. B.C.I. & P.: 456 (1960).—Harker & Napper, Ill. Guide Grass. Uganda: 20 (1960).—Napper, Grass. Tangan.: 15 (1965).—W. D. Clayton, F.T.E.A., Gramineae: 67, fig. 23 no. 14 (1970). TAB. **17** fig. B. Type from Ecuador.

Bromus catharticus Vahl, Symb. Bot. **2**: 22 (1791) *nomen confusum.*—Hack. in Engl. & Prantl, Nat. Pflanzenfam. **2**, 2: 75 (1887).—Monro in Proc. Rhod. Sci. Assoc. **6**, 1: 38, 61 (1906).—Chippindall in Meredith, Grasses & Pastures of S. Afr.: 63 (1955).—Type not certain.

Festuca unioloides Willd. in Hort. Berol. **1**: tab. 3 (1803). Type from North America (Carolina) cultivated from a seed at Berlin.

Ceratochloa unioloides (Willd.) Beauv., Ess. Agrost.: 75, 164, tab. 15, fig. 7 (1812). Type as above.

Schedonardus unioloides (Kunth) Roem. & Schult. in L., Syst. Veg. **2**: 708 (1817). Type as for *Bromus unioloides.*

Bromus unioloides (Willd.) Rasp. in Ann. Sci. Nat., Bot. **5**: 439 (1825) non Kunth (1816). Type as for *Festuca unioloides.*

Bromus willdenowii Kunth, Rev. Gram. **1**: 134 (1829). Type as for *Festuca unioloides.*

Ceratochloa haenkeana Presl in Reliq. Haenk. **1**: 285 (1830). Type from Chile.
Bromus haenkeanus (Presl) Kunth, Enum. Pl. **1**: 416 (1833). Type as above.
Ceratochloa cathartica (Vahl) Herter, Revist. Sudamer. Bot. **6**: 144 (1940). Type
as for *Bromus catharticus*.

A rather variable annual, biennial or short-lived perennial. Culms 40–100 cm.
tall, 2–4-noded, erect to straggling, slender and weak to robust, glabrous, smooth.
Leaf-sheaths tight at first later usually slipping off the culm, striate, inconspicuously
keeled; the lower usually densely pubescent to almost tomentose, rarely glabrous,
pallid. Ligule 2–5 mm. long, ovate, often lacerate. Leaf-laminae 6–25 × 0·2–1·2
cm., linear, tapering to a fine point, more or less flaccid, usually expanded, glabrous
or more rarely shortly pilose especially dorsally towards the base, scabrous on the
nerves of both surfaces and along the margins. Panicle 15–30 cm. long, usually
rather narrow, oblong, narrowly elliptic, or ovate-oblong in outline, erect or sometimes nodding, scantily branched (in poorly developed plants reduced to a raceme);
rhachis terete and grooved below, sharply angular above, increasingly scaberulous
towards the apex; branches 2–4-nate, remote, filiform, angulose, scabrous,
obliquely ascending to drooping. Spikelets (15)20–40 mm. long, 6–12-flowered,
pedicelled, narrowly elliptic to lanceolate-elliptic in outline, strongly laterally
compressed, pale green to glaucous; rhachilla easily disarticulating at maturity.
Glums lanceolate, with the apex acute to acuminate, with whitish hyaline margins,
glabrous, smooth but for the scaberulous keels; the inferior 6–14 mm. long,
5–7-nerved; the superior 11–17 mm. long, 7–9-nerved. Lemmas 12–18 mm. long,
9–13-nerved, very densely imbricate, narrowly oblong-lanceolate, chartaceous with
thinly membranous margins, with the apex usually very minutely 2-dentate,
mucronate or shortly awned, very sharply keeled, scaberulous along the flanks
(mainly on the nerves) and along the keel; awn up to 3 mm. long. Palea c. the
length of the lemma, with the keels stiffly ciliolate. Anthers 0·5–1 mm. long.
Caryopsis 8–12 mm. long.

Rhodesia. W: Matopos, 1400 m., viii. 1922, *Mainwaring* 3587 (SRGH). C: Marandellas, 19.x.1931, *Rattray* 5383 (BM; SRGH). E: Inyanga, 1710 m., iii.1931, *Allen* 3762
(SRGH).
Indigenous to the New World (S. America ?), introduced into Australia, Asia, southern
Europe, tropical and southern Africa where it usually grows as a weed. In S. Africa
some of the cultivated varieties are grown for pastures. In disturbed areas, in gardens
and arable land, often in moist places.

According to P. H. Raven (Brittonia, **12**: 219–221 (1960)) *B. unioloides* Kunth is
confined to the Andes whereas the widely disseminated grass should be referred to as
B. willdenowii Kunth. The main difference between these 2 taxa is supposed to lie in the
scabrousness of the nerves of the lemma: in the first sp. the nerves are said to be long-
ciliate with the intercostal spaces absolutely smooth whereas in the second sp. the flanks
of the lemmas should be " uniformly scabrous ". Raven also reports that the " 2 spp.
occupy distinct habitats and geographical areas in S. America. . . .". Although I am not
able to contest Raven's conclusions it was impossible for me to separate the 2 spp. in the
herbarium. In the same attempt my friends at the Kew Herbarium also failed (oral com-
munication by W. D. Clayton). Incidentally in no instance could we find " ciliate "
nerves on the lemmas unless the scabrous nerves were subjected to high magnifications.
There is a continuous range from the nerves being scabrous to entirely smooth. Unless
better evidence, based on more material, can be produced it seems advisable to maintain
the well-established name for the whole complex.

2. **Bromus leptocladus** Nees, Fl. Afr. Austr.: 453 (1841).—Dur. & Schinz, Consp. Fl.
Afr. **5**: 924 (1895).—Stapf in Harv. & Sond., F. C. **7**: 727, 731 (1900).—Wood,
Natal Pl. **5**, 3: t. 459 (1906).—Eggeling, Annot. List Grass. Uganda: 8 (1947).—
Chippindall in Meredith, Grasses & Pastures of S. Afr.: 65 (1955).—W. D. Clayton,
F.T.E.A., Gramineae: 68, fig. 23, 1–13 (1970). TAB. **17**, fig. A. Type from S.
Africa.
Bromus petitianus A. Rich., Tent. Fl. Abyss. **2**: 438 (1851). Type from Ethiopia.
Bromus cognatus Steud., Syn. Pl. Glum. **1**: 321 (1854).—Eggeling, loc. cit.—
Sturgeon in Rhod. Agric. Journ. **51**: 225 (1954).—C. E. Hubbard in Mem. N.Y.
Bot. Gard. **9**, 1: 102 (1954).—Jackson & Wiehe, Annot. Check List Nyasal. Grass.:
32 (1958).—Harker & Napper, Illustr. Guide Grass. Uganda: 21 (1960).—Napper,
Grass. Tangan.: 15 (1965). Type from Ethiopia.
Bromus scabridus Hook. f. in Journ. Linn. Soc., Bot. **7**: 231 (1864). Type from
the Cameroon Mts.

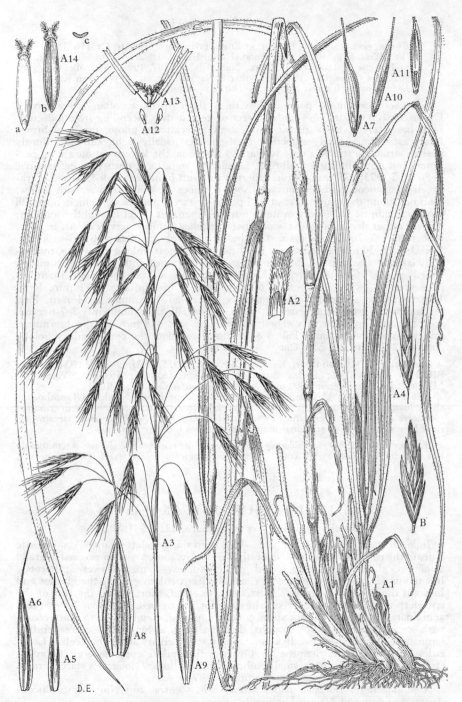

Tab. 17. A.—BROMUS LEPTOCLADUS. A1, habit ($\times\frac{1}{2}$); A2, junction of leaf-sheath and lamina showing ligule ($\times 1$), both from *Jackson* 1490; A3, inflorescence ($\times\frac{1}{2}$) *Robinson* 5268; A4, spikelet in lateral view ($\times 1$); A5, inferior glume ($\times 3$); A6, superior glume ($\times 3$); A7, inferior floret with extension of rhachilla ($\times 3$); A8, lemma ($\times 3$); A9, palea ($\times 3$); A10, second floret in lateral view ($\times 3$); A11, same from behind showing palea and insertion of rhachilla extension ($\times 3$); A12, lodicules ($\times 5$); A13, sexual organs ($\times 5$) all from *Lawrence* 435; A14, caryopsis (a) dorsal view, (b) ventral view, (c) cross-section (all $\times 5$) *Jackson* 2011. B.—BROMUS UNIOLOIDES, spikelet in lateral view ($\times 1$) *Whellan* 815.

Bromus runssoroensis K. Schum. in Engl., Pflanzenw. Ost-Afr. **C**: 116 (1895).—R.
E. Fr., Wiss. Ergebn. Schwed. Rhod.-Kong. Exped. **1**: 215 (1916).—Eggeling, loc.
cit.—Robyns & Tournay, Fl. Parc. Nat. Alb. **3**: 197, 199 (1955).—Bogdan, Rev.
List Kenya Grass.: 14 (1958).—Harker & Napper, loc. cit. Type from Kenya.
Bromus adöensis sensu Napper, loc. cit. non Steud. (1854).

A loosely caespitose perennial, with a short rhizome, often stoloniferous.
Culms 30–100 cm. tall, 2–4-noded, erect or geniculately ascending from a decumbent base, terete, stout, usually pubescent to scattered pilose, rarely glabrous,
finely asperulous, rarely smooth. Leaf-sheaths usually tight, tubular, bluntly
keeled, striate, scattered pilose to (rarely) glabrous, the lowermost ones long persistent but not splitting into fibres. Ligule 2–5 mm. long, obtuse. Leaf-laminae
8–30 × 0·5–1·25(1·75) cm., linear, tapering to a soft point, expanded, flaccid, dark-
green to glaucous, scaberulous on both surfaces or sometimes smooth below,
scabrous along the margins, scattered pilose or rarely glabrous. Panicle up to 28
cm. long, elliptic to broadly ovate in outline, open, lax, erect to slightly nodding;
rhachis rather slender, angular, scabrous; branches filiform, very flexuous, angular,
scaberulous, 2–3-nate, not divided in the lower $\frac{1}{2}$–$\frac{1}{3}$; branchlets with 1–3-spikelets.
Spikelets 15–30 mm. long, loosely 5–12-flowered, linear-oblong in outline, moder-
ately laterally compressed, light green, sometimes tinged with purple, glabrous or
shortly pilose. Glumes unequal, broadly lanceolate to ovate-lanceolate, membra-
nous-scarious, with the apex acute to subacute; the inferior 5–11 mm. long,
1-nerved; the superior 7–14 mm. long, 3-nerved. Lemmas 5–14 mm. long
(excluding the awn), lanceolate-elliptic to oblong-elliptic, prominently 3–7-nerved,
with the apex entire or very shortly 2-dentate; awn 4–8 mm. long, subterminal,
fine, straight, scaberulous. Paleas 8–11 mm. long, with the keels rigidly ciliolate.
Anthers 2–3·75 mm. long. Caryopsis c. 8 mm. long.

Rhodesia. E: Umtali Distr., Engwa, 1830 m., 1.ii.1955, *E. M. & W.* 81 (BM; LISC;
SRGH). **Malawi.** C: Dedza 6.ix.1956, *Jackson* 163 (K; SRGH; NYAS). S:
Mlanje, Lichenya Plateau, 1800 m., 6.vi.1962, *Robinson* 5268 (K; SRGH).
Also recorded from Uganda, Kenya, Tanzania, Cape Prov., Natal and Lesotho. A
shade-loving species usually in moist ground in forest patches, on edges of riparian
rain-forests, in montane evergreen forest, in secondary scrub, along roads or streams, on
river banks, wet vleis or similar situations, sometimes locally common.

B. leptocladus is a very variable grass; in particular the range of spikelet size is remarkable
but since the variation is quite continuous no distinct infraspecific taxa can be recognised.

IX. BRACHYPODIEAE Harz

Brachypodieae Harz in Linnaea, **43**: 15 (1880).

Inflorescence a raceme, often spike-like, with the spikelets arranged on opposite
sides of the rhachis and facing it laterally. Spikelets somewhat compressed laterally
to almost cylindrical, all alike, shortly pedicelled, few- to many-flowered; florets ☿
but the uppermost often reduced; rhachilla disarticulating above the glumes and
beneath the florets. Glumes 2, persistent, unequal, shorter than the rest of the
spikelet; the inferior 1–5-nerved; the superior 3–7-nerved. Lemmas 7–9-nerved,
acuminate and awned from the apex, dorsally rounded, membranous to chartaceous,
with scarious margins; awn straight. Paleas usually as long as the corresponding
lemmas, 2-nerved, 2-keeled, hyaline, usually stiffly ciliate along the keels. Lodicules 2, often ciliate. Stamens 3. Ovary with a pilose appendage at the apex;
styles 2. Caryopsis with a small embryo; hilum linear, elongate; starch grains
simple. Chromosomes small, basic number 5, 7 or 9.
A monogeneric Eurasian tribe, but also in Central and North (introduced)
America and on the African Highlands.

16. BRACHYPODIUM Beauv.

Brachypodium Beauv., Agrost.: 100 (1812).

Spikelets subsessile, solitary, several- to many-flowered, at first cylindric, later
elatrally slightly compressed, usually awned. Florets ☿, all fertile but the upper-

most usually reduced. Rhachilla disarticulating above the glumes and between the florets, glabrous, smooth. Glumes unequal, 3–7(–9)-nerved. Lemmas 7–9-nerved, oblong to oblong-lanceolate, densely imbricate at first, later somewhat spreading, mucronate or with a straight awn, dorsally rounded: callus very short, obtuse. Paleas slightly shorter than the corresponding lemmas, oblong, with the apex broadly obtuse to truncate, 2-keeled, pectinate-ciliate along the keels. Lodicules 2, lanceolate, with the apex usually truncate, ciliate or more rarely glabrous. Stamens 3, rarely 2. Ovary with a villous apical appendage or simply hairy; styles inserted at the appendage, short; stigmas plumose, laterally exserted. Caryopsis oblong to linear in outline, convex-concave, with an apical ciliate appendage, usually tightly enclosed by the palea; hilum filiform, very long. Inflorescence a simple terminal raceme (pseudo-spike), with the spikelets alternating in 2 rows on opposite sides of the rhachis. Perennial or more rarely annual grasses of varying habits, often rhizomatous. Ligule membranous.

A predominantly Eurasian genus of c. 15 spp., also introduced into North America, 2 or 3 spp. in temperate regions of N. and S. Africa and on mountains in tropical Africa.

Brachypodium flexum Nees, Fl. Afr. Austr.: 456 (1841).—Engl., Pflanzenw. Ost-Afr. **C**: 116 (1895);—op. cit. **A**: 126, 130 (1895).—Stapf in Harv. & Sond., F. C. **7**: 735 (1900).—Wood, Natal Pl. **5**, 3: t. 462 (1906).—Sturgeon, Rev. List Grass. S. Rhod. in Rhod. Agric. Journ. **51**: 225 (1954).—W. D. Clayton in F.T.E.A., Gramineae: 71, fig. 24 (1970). TAB. **18**. Type from S. Africa.

    Brachypodium flexum var. *abyssinicum* Schimper, Iter Abyss. **2**: 674 (1842).—Robyns & Tournay, Fl. Parc. Nat. Alb. **3**: 200 (1955).—F. W. Andr., Fl. Pl. Sudan, **3**: 413 (1956). Type from Ethiopia.

    Triticum flexum (Nees) A. Rich., Tent. Fl. Abyss. **2**: 441 (1851). Type as for *Brachypodium flexum*.

    Festuca diaphana Steud., Syn. Pl. Glum. **1**: 316 (1854) *nom. superfl.*

    Brachypodium multiflorum Engl. in Abh. Preuss. Akad. Wiss. Berl. **1894**: 58 (1894) *nom. nud.*, non Bercht. (1836).

    Dinebra pubescens K. Schum. in Engl., Pflanzenw. Ost-Afr. **C**: 111 (1895). Type from Tanzania.

    Brachypodium pubescens Engl., Pflanzenw. Ost-Afr. **A**: 128 (1895) *nom. nud.*

    Brachypodium schumannianum Pilg. in Notizbl. Bot. Gart. Berl. **9**: 1135 (1927). Type from Tanzania.

    Brachypodium diaphanum Cufod. in Bull. Jard. Bot. Brux., Suppl., **38**: 1215 (1968). Type from Ethiopia.

A very variable perennial. Culms up to 100 cm. tall, 5–many-noded, usually branched below, rarely simple, straggling, ascending from a geniculate base, or often decumbent and rooting from the lower nodes, terete, wiry, rather slender, smooth, glabrous or sometimes hairy below the nodes; nodes finely pubescent. Leaf-sheaths shorter than the internodes, striate, tight at first, later slipping off the culm, with a row of spreading hairs along the margins, otherwise glabrous or rarely scattered pilose. Ligule c. 1 mm. long, truncate. Leaf-laminae 5–15 × 0·3–0·8 cm., linear, long tapering to a fine flexible point, dark-green to glaucous, somewhat rigid to almost flaccid, usually scattered pilose, sometimes glabrous, smooth on the upper, scaberulous on the lower surface. Raceme 4–15 cm. long, erect or more often pendulous, bearing 3–9 spikelets; rhachis very slender, frequently curved, compressed, glabrous, scabrous mainly along the margins. Pedicels hardly more than 1·5 mm. long. Spikelets 1·25–4·5 cm. long, 7–many-flowered, usually overlapping but sometimes remote, horizontally spreading to obliquely ascending, glabrous or thinly pubescent. Glumes with the apex acute to acuminate, chartaceous, strongly nerved; inferior 3–6 mm. long, (3)4–5-nerved, narrowly triangular to lanceolate-subulate; superior 5–8·5 mm. long, 7-nerved, lanceolate to lanceolate-oblong. Lemmas 6–8 mm. long, 7-nerved, firmly membranous, lanceolate, tapering into the awn, inconspicuously scaberulous, rarely shortly pubescent; awn 3–7·5 mm. long, slender, straight. Paleas 7–8 mm. long, with rigid cilia along the keels. Anthers c. 3 mm. long, linear.

Zambia. E: Nyika Plateau, 2130 m., 2.i.1959, *Robinson* 3003 (K; SRGH). **Rhodesia.** E: Inyanga, Troutbeck, 21.iii.1948, *Rattray* 1432 (K; SRGH). **Malawi.** N: Mzimba, Vipya, 17.iii.1957, *Jackson* 430 (K; SRGH). C: Dedza, 6.ix.1950, *Jackson* 164 (K; SRGH). S: Mlanje, Chambe basin, 1800 m., 14.vi.1962, *Robinson* 5342 (SRGH). **Mozambique.** T: Angónia, on the side of Mt. Dómuè, 1500 m., 16.x.1943, *Torre* 6050

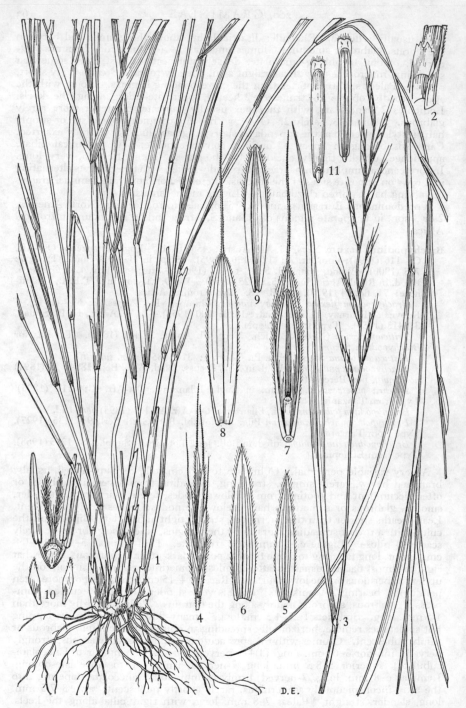

Tab. 18. BRACHYPODIUM FLEXUM. 1, habit (× ⅔); 2, ligule (× 4), both from *Robinson* 3003; 3, inflorescence (× ⅔) *Jackson* 430; 4, spikelet (× 2); 5, inferior glume (× 4); 6 superior glume (× 4); 7, floret (× 4); 8, lemma (× 4); 9, palea (× 4) all from *Robinson* 3003; 10, sexual organs with lodicules detached (× 6) *Jackson* 164; 11, caryopsis ventral and dorsal views (× 4) *Robinson* 3003.

(LISC). MS: Manica, Mt. Zuira, Tsetsera, c. 1800 m., 4.iv.1966, *Torre & Correia* 15690 (LISC).

In the southern and eastern Cape Prov., Lesotho, Natal, Zululand, Transvaal, Tanzania, Uganda, Kenya, Ethiopia and the Sudan. In open forests, forest outskirts, along forest margins and in moist shady localities such as stream-sides and river banks, often rambling over bushes.

B. flexum very closely resembles *B. mexicanum* (Roem. & Schult.) Link. In that sp., however, the glumes are of a tougher consistency with the nerves dorsally less prominent, and, moreover, the apex of the glumes varies from being acute to subobtuse whereas it is almost always obliquely truncate in *B. flexum*. It is also closely related to the European *B. sylvaticum* (Huds.) Beauv. from which sp. it can be distinguished only by the scabridity of both culms and leaf-sheaths.

X. TRITICEAE Dumort.

Triticeae Dumort, Obs. Gram. Belg.: 82 (1823).

Inflorescence a spike, erect or drooping. Spikelets 1–many-flowered, arranged in groups of 2–3 alternately on opposite sides of the rhachis, all sessile and alike (♀) or the lateral ones of a triad pedicelled, ♂ or barren, or much reduced, usually laterally compressed; rhachilla disarticulating above the glumes and beneath each floret in wild species, not disarticulating in cultivated species. Glumes persistent, as long as or shorter than the spikelet, usually coriaceous, strongly nerved, usually well developed, rarely reduced, occasionally rather narrow. Lemmas 5–9-nerved, awnless or awned, usually coriaceous; awn usually apical, straight or curved but never geniculate or twisted. Paleas about as long as the lemmas. Lodicules 2. Stamens 3. Ovary crowned by a pilose appendage; stigmas 2, plumose. Caryopsis often adhering to lemma and palea; embryo up to $\frac{1}{2}$ as long as the caryopsis but usually smaller, pooid; hilum linear, oblong; starch grains simple. Basic chromosome number 7.

A tribe of about 10 genera, containing the most important temperate cereals as well as a large number of weeds.

None of these cereals is of great economic importance in countries covered by the Flora Zambesiaca area. In many instances their cultivation is still in the experimental stage. Three genera are mainly to be considered; they can be keyed out as follows:

Spikelets 3 at each node of the rhachis, the 2 lateral ones pedicelled   -   **Hordeum**
Spikelets always solitary at each node of the rhachis:
    Spikelets 2-flowered. Glumes 1-nerved, subulate. Lemmas on the keels stiffly pilose
    or pectinate-ciliate -   -   -   -   -   -   -   -   -   **Secale**
    Spikelets 3–many-flowered. Glumes with several nerves usually ovate or oblong.
    Lemmas with the keels neither stiffly pilose nor pectinate-ciliate - **Triticum**

Triticum aestivum L. (wheat) is grown fairly extensively in the SE. lowveld of Rhodesia. In Zambia it is cultivated on a small scale (1969 c. 250 acres) by a few commercial producers and Government sponsored enterprise. In the same year c. 1000 acres were grown as a winter crop under irrigation in Botswana. The cultivation of wheat in Mozambique goes back a long time and quite a number of varieties are well established. For more precise information the following literature should be consulted:

Barradas, L. A cultura do trigo no Sul do Save, in Gazeta do Agricultor, No. 13. June 1950.
—— Preparemo-nos para a próxima campanha do trigo, op. cit. No. 21. February 1951.
Barreiros, H. A. A cultura do trigo (conclusão) práticas dos ensaios realizados na Manhica, in Gazeta do Agricultor, No. 35. April 1952.
Faleão, A. H. de Sousa. A experimentacão no Posto de Culturas regadas do Vale de Limpopo. Estudos, Ensaios e Documentos No. 33, Junta de Investigações do Ultramar. Lisbon 1957.

c

Ferreira, A. Campanha do trigo in Gazeta do Agricultor, No. 58. March 1954.
Silva, R. N. O problema do trigo em Moçambique, in Gazeta do Agricultor,
     No. 6. November 1949.
Trigo em Moçambique—Noticia da secção Agronómica dos Serviços de Agri-
     cultura, in Gazeta do Agricultor, No. 57. February 1954.
Hordeum vulgare L. (barley) is grown on a small scale in both Rhodesia and
Mozambique (see F. Santos Viana, Breves notas sobre a cultura da cevada em
Moçambique in Gazeta do Agricultor, No. 98. July 1957), but not at all in Bot-
swana. In Zambia a small acreage of barley was put in on a Government farm in
the 1969–1970 season.
Secale cereale L. (rye) has been grown experimentally in Zambia and Mozam-
bique but apparently with little success.
The above information was kindly supplied by the Chief Research Officer,
Dept. of Agriculture of Zambia; by the Director of Agriculture of Botswana as well
as by B. K. Simon of Salisbury, Rhodesia and M. Myre of Mozambique. Further
literature on the subject of cultivated *Triticeae* is found in Clayton, in F.T.E.A.,
Gramineae: 73 (1970).

XI. MELICEAE Reichenb.

Meliceae Reichenb., Consp. Reg. Veg.: 53 (1828) as Melicaceae.—
          Tateoka in Bot. Mag. Tokyo, **78**: 289 (1965).

Inflorescence a panicle, often contracted and then spike-like. Spikelets laterally
compressed, 1–many-flowered, the lower florets ☿, the uppermost reduced and
often only represented by clusters of small empty lemmas; rhachilla articulating
above the glumes and below each floret, with long internodes, often terminating
into a clavate appendage. Glumes 2, persistent, equal or unequal, membranous,
usually shorter than the rest of the spikelet; the inferior 1–3-nerved; the superior
5-nerved. Lemmas 5–9-nerved, entire or 2-dentate, awnless or awned from just
below the apex, membranous or somewhat chartaceous, with hyaline margins.
Paleas slightly shorter than the corresponding lemmas, 2-nerved, 2-keeled, hyaline.
Lodicules 2, rather fleshy, truncate, often laterally fused. Stamens 3. Ovary
glabrous; styles 2. Caryopsis fusiform or terete; embryo small; hilum linear;
starch grains compound. Chromosomes large, basic number 9.
A tribe of c. 5 genera, throughout the temperate regions of the world but some
spp. also extending into tropical areas.

17. STREBLOCHAETE Hochst. ex Pilg.

Streblochaete Hochst. ex Pilg. in Engl., Bot. Jahrb. **37**, Beibl. no. 85: 61 (1906).

Spikelets pedicelled, (1)2–6-flowered, laterally compressed, linear-lanceolate
to narrowly oblong in outline, long awned; florets (the lowermost spikelet often
1-flowered) loosely arranged, the lower 2–4 usually bisexual, the following either
male or greatly reduced; callus relatively long, slender, linear, densely whitish-
hairy; rhachilla glabrous, disarticulating above the glumes and between the
florets, usually produced beyond the uppermost floret. Glumes 2, subequal,
persistent, apex subacute to acute, membranous, with hyaline margins; the inferior
slightly smaller than the superior, 3-nerved; the superior 5-nerved, often with
transverse veinlets. Lemmas much exceeding the glumes, distinctly 7-nerved,
expanded lanceolate to lanceolate-elliptic or narrowly elliptic, herbaceous at first,
later chartaceous, apex shortly bilobed or more rarely entire, awned from the back
beneath the sinus of the lobes; awn rather long, increasingly slender towards the
apex, scaberulous, twisted, becoming entangled with the awns from the other
florets. Paleas shorter than the lemmas, 2-keeled, linear, with thin hyaline
margins. Lodicules 2, minute, cuneate-truncate, fleshy, glabrous. Stamens 3;
filaments broadened towards the base; anthers linear-oblong. Ovary glabrous,
oblong in outline; styles distinct; stigmas plumose, laterally exserted. Caryopsis

linear-oblong in outline, almost terete, dorsally with a long terminal groove, the apex crowned with the persistent base of the style; embryo small; hilum minute, basal. Perennial grasses with flaccid leaves. Ligule a short membrane. Inflorescence an open somewhat secund panicle.

A monospecific genus, from the tropical and subtropical mountains of Africa, also recorded from Réunion and the Philippine Is.

Streblochaete longiarista (A. Rich.) Pilg. in Notizbl. Bot. Gart. Berl. **9**: 516 (1926).— Peter in Fedde Repert., Beih. **40**: 305 (1931).—C. E. Hubb. in F.W.T.A. **2**: 528 (1936).—in Kew Bull. **1936**: 330 (1936).—in F.T.A. **10**: 102 (1937).—Eggeling, Annot. List Grass. Uganda: 46 (1947).—Robyns & Tournay in Fl. Parc Nat. Alb. **3**: 180 (1955).—Chippindall in Meredith, Grasses & Pastures of S. Afr.: 82 (1955). —Jackson & Wiehe, Annot. Check List Nyasal. Grass.: 62 (1958).—Bogdan, Rev. List Kenya Grass.: 30 (1958).—Napper, Grass. Tangan.: 17 (1965).—W. D. Clayton, Key Nig. Grass., ed. 2: 9 (1966).—W. D. Clayton in F.T.E.A., Gramineae: 74 fig. 25 (1970). TAB. **19**. Type from Ethiopia.

    Trisetum longiaristum A. Rich., Tent. Fl. Abyss. **2**: 417 (1851). Type as above.

    Bromus trichopodus A. Rich., tom. cit.: 437 (1851). Type from Ethiopia.

    Streblochaete nutans Hochst. ex A. Rich., tom. cit.: 417 (1851). Type from Ethiopia.

    Danthonia streblochaeta Steud., Syn. Pl. Glum. **1**: 245 (1854).—Hook. f. in Journ. Linn. Soc., Bot. **7**: 229 (1864).—K. Schum. in Engl., Pflanzenw. Ost-Afr. **C**: 110 (1895). Type from Ethiopia.

    Danthonia longiarista (A. Rich) Engl. in Abh. Preuss. Akad. Wiss. **1891**: 130 (1892). Type as for *Streblochaete longiarista*.

A loosely caespitose straggling perennial. Culms up to 90 cm. tall, 3–8-noded, simple or more rarely branched, terete, rather slender, geniculately ascending, glabrous, smooth. Leaf-sheaths shorter than the internodes, striate, somewhat loose, keeled in the upper half, scabrous along the nerves or smooth, glabrous or rarely shortly pilose, with the mouth auricled (the auricles erect, adnate to the ligule). Ligule 6–12 mm. long, often lacerate. Leaf-laminae 7–27 × 0·3–1·1 cm., narrowly linear from an attenuate base, long-tapering to an acute point, flaccid, usually drooping, usually expanded, vivid-green, scaberulous, usually glabrous, rarely pubescent on the upper surface. Panicle 10–17(20) cm. long, rather lax, contracted; rhachis slender, smooth or scaberulous towards the apex; branches up to 10 cm. long becoming gradually shorter towards the apex, filiform, solitary or (the lower) 2–3-nate, the lower ones usually divided the upper simple. Spikelets 15–27 mm. long, pale green to silvery-green; rhachilla internodes 2–3-mm. long, slender, glabrous; callus 2–3 mm. long. Inferior glume 6–10 mm. long, linear-lanceolate; the superior 10–14 mm. long, oblong-lanceolate. Lemmas 13–16 mm. long, becoming gradually shorter towards the apex, glabrous, with the apical lobes 1–2 mm. long, acute; awn 20–43 mm. long. Paleas up to 10 mm. long, the keels usually scaberulous towards the apex. Anthers 2·5–3 mm. long. Caryopsis c. 5 mm. long.

Rhodesia. E: Melsetter Sub-Station, 22.v.1953, *West* 3323 (K; SRGH). **Malawi.** N: Vipya, 10.vii.1952, *Jackson* 963 (K; SRGH). C: Dedza Mt., 6.ix.1950, *Jackson* 157 (K; SRGH).

Also in Nigeria, Cameroun, Ethiopia, Uganda, Kenya, Tanzania, the eastern Cape Prov. and Natal. In montane or submontane evergreen forest scrub.

This plant has an interesting method of seed dispersal. By the intertwined hygroscopic awns the florets become detached from the panicle and are dispersed *en masse*.

XII. AVENEAE Dumort.

Aveneae Dumort., Obs. Gram. Belg.: 82, 120 (1823).

Inflorescence an open or sometimes contracted panicle, rarely a raceme. Spikelets all alike, awned, 2–7-flowered, moderately laterally compressed, pedicelled; florets usually all ♀ with the uppermost barren or rarely the 1st and 2nd ♂ or barren, if only 2 then sometimes the inferior ♂ and the superior ♀; rhachilla

Tab. 19. STREBLOCHAETE LONGIARISTA. 1, habit (×⅔); 2, junction of leaf-sheath and
lamina showing ligule (×2), both from *Wiehe* 640; 3, spikelet in lateral view (×2);
4, spikelet with the glumes removed showing 2 florets, callus and rhachilla internodes
(×4); 5, inferior glume (×2); 6, superior glume (×2); 7, lemma (×2); 8, palea
(×2); 9, sexual organs with lodicules detached (×2); 10, caryopsis in dorsal and
lateral views (×4) all from *Jackson* 521.

disarticulating above the glumes and usually between the florets (not disarticulating in the cultivated oats). Glumes 2, persistent, usually subequal but
sometimes the inferior shorter, as long as or longer than the spikelet (shorter
than the spikelet in some spp. of *Helictotrichon*), membranous to chartaceous,
often with shiny silvery margins. Lemmas 5–7-nerved, membranous to cartilaginous, with the margin hyaline or scarious, apex entire or 2(–4)-lobed (the lobes
often extended into bristles), dorsally awned, rarely awnless; awns geniculate
rarely straight, with a usually twisted subterete column. Palea as long as the lemma,
rarely shorter, 2-nerved, hyaline. Lodicules 2, lanceolate, often 2-lobed, rarely
absent. Stamens 1–3. Ovary glabrous or pilose; stigmas with the styles plumose.
Caryopsis with small embryo; hilum linear. Starch grains compound. Chromosomes large; the basic number is 7.

A tribe of c. 25 genera distributed throughout the temperate regions of both
hemispheres, extending into mountainous areas of the tropics.

Lemmas awnless, sometimes mucronate. Rhachis of the (usually contracted) panicle
  conspicuously tomentose or densely pubescent - - - - - 18. **Koeleria**
Lemmas always awned. Rhachis of the panicle glabrous or scabrous, if pilose then
  neither tomentose nor densely pubescent:
Spikelets strictly 2-flowered, 2–4·5 mm. long, with both florets ♀. Annual plants
              22. **Aira**
Spikelets at least 2–6-flowered, always longer than 5·5 mm. Perennial or annual plants:
  Spikelets with 3 florets; the 1st and 2nd barren, reduced to dorsally awned lemmas;
    the 3rd fertile, ♀, enveloped by the preceeding barren ones. 19. **Anthoxanthum**
  Spikelets with 2–6 florets (the 1st 3 reduced), with all the lemmas awned:
    Glumes 7–11-nerved, often slightly cross-veined. Weak annuals - 20. **Avena**
    Glumes 1–3-nerved, never cross-veined. Perennials. - - 21. **Helictotrichon**

18. KOELERIA Pers.

Koeleria Pers., Syn. Pl. **1**: 97 (1805).

Spikelets very shortly pedicelled, laterally compressed, (rarely 1) 2–many-
flowered, muticous or awned; florets ♀ or the uppermost one more or less reduced;
rhachilla disarticulating above the glumes and between the florets, produced or
often terminated by a reduced lemma, pubescent or glabrous. Glumes 2, unequal
to subequal, persistent, subacute to acuminate or more rarely obtuse, with hyaline
margins; the inferior 1–3-nerved; the superior 3–5-nerved. Lemmas almost
always exceeding the glumes, 3–5 (rarely 7–)-nerved, the lateral nerves usually not
very distinct, the middle nerve prominent and sometimes excurrent into a short
mucro, both margins and apex hyaline; callus very short, glabrous or shortly pilose.
Paleas shorter than or almost as long as the corresponding lemmas, 2-keeled,
hyaline, with the apex 2-dentate. Lodicules 2, very small, hyaline. Stamens 3.
Ovary glabrous; styles distinct, rather short; stigmas plumose, laterally exserted.
Caryopsis linear-oblong to oblong in outline, laterally compressed, of soft consistency, tightly embraced by the lemma; hilum basal, very minute. Perennials
or annuals of varying habitats. Ligule a hyaline membrane. Inflorescence a
panicle, usually more or less contracted, rarely lax and open, usually almost
cylindrical, often interrupted in the lower part, glabrous or hairy.

A genus of c. 3 not well defined spp., distributed throughout the temperate areas
of the globe, also in montane regions of tropical and subtropical Africa on both sides
of the equator.

Koeleria capensis (Steud.) Nees in Linnaea, **7**: 321 (1832).—W. D. Clayton in F.T.E.A.,
  Gramineae: 79, fig. 27 (1970). TAB. **20**. Type from S. Africa.
  Aira capensis Steud. in Flora, **12**: 469 (1829). Type as above.
  Koeleria cristata sensu A. Rich., Tent. Fl. Abyss. **2**: 431 (1851).—Hook. f. in
  Journ. Linn. Soc., Bot. **7**: 230 (1864).—Hack. in Engl. & Prantl, Nat. Pflanzenfam.
  2, 2: 70 (1887).—Engl. in Hochgebirgsfl. Trop. Afr.: 134 (1892) pro parte.—
  Dur. & Schinz, Consp. Fl. Afr.: 892 (1894) pro parte.—Engl., Pflanzenw. Ost-Afr.
  C: 115 (1895) pro parte.—Stapf in Harv. & Sond., F.C. **7**: 468 (1899).—Monro
  in Proc. Rhod. Sci. Assoc. **6**, 1: 45 (1906).—Sturgeon in Rhod. Agric. Journ. **51**:
  223 (1954).—Chippindall in Meredith, Grasses & Pastures of S. Afr.: 83 (1955).—
  Jackson & Wiehe, Annot. Check List Nyasal. Grass.: 46 (1958).
  Koeleria convoluta Steud., Syn. Pl. Glum. **1**: 293 (1854).—Domin in Biblioth. Bot.
  65: 109 (1907).—Chiov. in Ann. Ist. Bot. Roma, **8**: 374 (1908).—Pilg. in Notizbl.

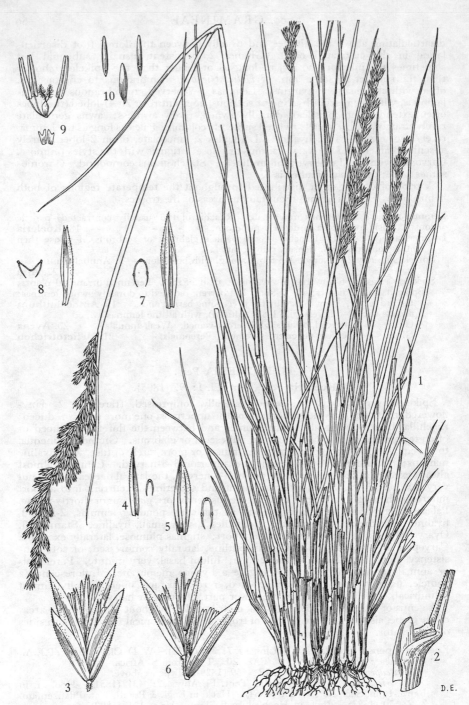

Tab. 20. KOELERIA CAPENSIS. 1, habit (×⅔) *Wiehe* N310 and N360; 2, ligule (×4);
3, spikelet (×4); 4, inferior glume (×4) with transverse section (×6); 5, superior
glume (×4) with transverse section (×6); 6, spikelet with glumes removed (×4);
7, lemma (×4) with transverse section (×6); 8, palea (×4) with transverse section
(×6); 9, sexual organs with lodicules detached (×4), all from *Wiehe* N360; 10,
caryopsis, dorsal, ventral and lateral views (×4) *Greenway* 6291.

Bot. Gart. Berl. **9**: 1126 (1927).—Hutch., F.W.T.A. **2**: 527 (1932). Type from Ethiopia.

Airochloa uniflora Hochst. in Flora, **38**: 330 (1855).—Type from Ethiopia.

Koeleria uniflora (Hochst.) Schweinf. & Aschers. in Schweinf., Beitr. Fl. Aethiop.: 300 (1867). Type as above.

Koeleria convoluta var. *typica* Domin, tom. cit.: 110 (1907). Type from Ethiopia.

Koeleria convoluta var. *uniflora* (Hochst.) Domin, loc. cit. Type as for *Airochloa uniflora*.

Koeleria convoluta var. *densiflora* Domin, tom. cit.: 111 (1907). Type from Ethiopia.

Koeleria convoluta var. *deusticola* Domin, loc. cit. Type from Tanzania.

Koeleria convoluta subvar. *supina* Domin, tom. cit.: 112 (1907). Type from Tanzania.

Koeleria wildemanii Domin, loc. cit. Type: Malawi, Mt. Mlanje, *Clounie* 64 (K, holotype).

Koeleria cristata var. *densiflora* (Domin) Peter, Fl. Deutsch Ost-Afr. **1**: 342 (1936). Type as for *Koeleria convoluta* var. *densiflora*.

Koeleria cristata var. *shiraensis* C. E. Hubb. in Kew Bull. **1936**: 500 (1936). Type from Tanzania.

Koeleria cristata var. *convoluta* (Steud.) C. E. Hubb. in F.T.A. **10**, 1: 95 (1937). Type as for *Koeleria convoluta*.

Koeleria cristata var. *brevifolia* (Nees) C. E. Hubb., tom. cit.: 96 (1937). Type from S. Africa.

Koeleria cristata var. *supina* (Domin) C. E. Hubb., tom. cit.: 97 (1937).—Napper, Grass. Tangan.: 17 (1965). Type as for *Koeleria convoluta* subvar. *supina*.

Koeleria gracilis var. *convoluta* (Steud.) Hedberg in Symb. Bot. Uppsal. **15**: 42 (1957). Type as for *Koeleria convoluta*.

Koeleria gracilis var. *supina* (Domin) Hedberg, Afroalp. Vasc. Pl.: 42 (1957). Type as for *Koeleria convoluta* subvar. *supina*.

Koeleria pyramidata var. *brevifolia* (Nees) Cufod. in Bull. Jard. Bot. Brux., Suppl., **38**: 1227 (1968). Type as for *Koeleria cristata* var. *brevifolia*.

Koeleria pyramidata var. *convoluta* (Steud.) Cufod., loc. cit. Type as for *Koeleria convoluta*.

A densely caespitose very variable perennial. Culms 15–90 cm. tall, 1–2-noded, erect or rarely geniculately ascending, simple, slender, almost wiry, terete, glabrous to sometimes thinly tomentose in the upper part. Leaves densely crowded at the slightly thickened base; leaf-sheaths longer than the lower internodes, striate, usually somewhat loose, often papery, long persistent, glabrous or pubescent. Ligule c. 2 mm. long. Leaf-laminae 3–35 cm. long, very narrowly linear to filiform, tapering to a hard bluntish tip, usually convolute or simply plicate, somewhat stiff, usually erect or nearly so, glabrous or pubescent, smooth or finely asperulous. Panicle (3)5–12(17) cm. long, erect, more or less cylindric, rather condensed, almost spike-like, continuous, or sometimes lobed and often interrupted; rhachis pubescent to thinly tomentose. Spikelets 4–7 mm. long, 2–4-flowered, elliptic to elliptic-obovate in lateral view, gaping, green, often tinged with purple, or sometimes variegated with brown and purple. Glumes with the keels scabrous or sometimes ciliolate towards the apex, margins hyaline and thin; the inferior 3·5–6 mm. long, 1-nerved, narrowly lanceolate; the superior 4·5–6 mm. long, usually 3-nerved, elliptic, apex acuminate. Lemmas 4–6·5 mm. long, 3-nerved, elliptic to elliptic-oblong, apex acuminate, with the keel scaberulous or sometimes ciliolate, the flanks asperulous or finely puberulous. Anthers 2–3 mm. long, linear. Caryopsis c. 3 mm. long.

Zambia. E: Lundazi, Nyika Plateau, ix.1968, *Williamson* 1015 (BM; K). **Rhodesia.** E: Umtali, N. boundary Sheba Farm, 1680 m., 23.ix.1952, *Chase* 4653 (BM; K; PRE; SRGH). **Malawi.** N: Nyika Plateau, on hill slopes, 22.ii.1949, *Hawksworth* in GHS 22685 (K; SRGH). S: Mt. Mlanje, 1680 m., 4.ix.1956, *Newman & Whitmore* 664 (SRGH). **Mozambique.** MS: Manica, Rotanda, Mts. of Messambize, 1600 m., 26.xi. 1965, *Torre & Correia* 13309 (LISC).

In Cameroun, Ethiopia, Uganda, Kenya, Tanzania and southwards to the Cape Prov. of S. Africa. In open high-altitude grassland, on acid humic soil, often on rocks and steep slopes, often dominant.

Koeleria capensis cannot easily be distinguished from the European *K. cristata* by spikelet characters, but vegetatively it can easily be recognised by the decaying broad soft papery leaf-sheaths. In *K. cristata* the inflorescence is usually less contracted than in the African spp.

19. ANTHOXANTHUM L.

Anthoxanthum L., Sp. Pl. 1: 28 (1753); Gen. Pl., ed. 5: 17 (1754).

Spikelets solitary, pedicelled, moderately laterally compressed, usually 3-flowered (rarely less), awned; rhachilla disarticulating above the superior glume, not produced beyond the uppermost floret. Florets heteromorphous, the first and second sterile or male (sometimes the first male and the second sterile), the terminal one always bisexual. Glumes 2, unequal, persistent, keeled, membranous, 1–3-nerved with the nerves usually prominent, apex acute to acuminate; the inferior about $\frac{3}{5}$–$\frac{3}{4}$ the length of the superior; the superior as long as or longer than the body of the lemmas. The first and second lemmas similar and almost equal, membranous, narrowly oblong in profile, 5–7-nerved, strongly laterally compressed, keeled, densely pilose, 2-fid to deeply 2-lobed, awned from the base or from the back or from the sinus; the awns usually unequal, that of the first lemma usually short and inserted $\frac{1}{3}$–$\frac{1}{2}$-way below the apex, usually straight; that of the second lemma as long as or longer than its body, inserted near the base or sometimes around the middle, geniculate. Terminal lemma muticous, much shorter than ($\frac{1}{3}$–$\frac{1}{2}$) the preceding ones, expanded broadly elliptic or ovate-elliptic, faintly 7–1-nerved, obtusely 2-keeled, thinly membranous. Paleas of the first and second floret, when present, slightly shorter than the lemmas, linear, hyaline, 2-keeled; the one of the terminal floret narrowly oblong, 2-keeled, usually 1-nerved. Lodicules absent. Stamens 3 in the male, 2 in the bisexual florets. Ovary glabrous; styles distinct, long; stigmas long, rather slender, plumose, exserted from the apex. Caryopsis oblong-elliptic in outline, slightly laterally compressed; embryo c. $\frac{1}{4}$ the length of the caryopsis; hilum small, punctiform, basal. Sweet-scented annuals or perennials, often stoloniferous. Ligules membranous. Inflorescence a dense spike-like panicle.

A genus of c. 20 spp., indigenous to the holarctic region, with 4 spp. in S. Africa, one of them extending into southern tropical Africa, and one species indigenous in the east African mountains.

Anthoxanthum ecklonii (Nees) Stapf in Harv. & Sond., F.C., 7: 466 (1900). TAB. 21.
　　Type from the Cape Prov. of S. Africa.
　　　　Ataxia ecklonis Nees ex Trin. in Mém. Acad. Sci. Petersb., Sér. 6, 5: 77 (1839).—
　　Steud., Syn. Pl. Glum. 1: 13 (1854). Type as above.
　　　　Hierochloa ecklonii Nees, Fl. Afr. Austr.: 7 (1841). Type as above.

A strongly aromatic rhizomatous perennial. Culms 12–75 cm. tall, 2–4-noded, erect or ascending from a decumbent base, simple or rarely branched from the lower nodes, terete, weak, rather slender, smooth, glabrous. Leaf-sheaths shorter than the internodes, rather tight at first, striate, smooth or reversedly scaberulous; the basal ones often with the laminas reduced to scales. Ligules 2–5 mm. long, scarious, whitish, obtuse, fringed with short cilia. Leaf-laminae 1–15 × 0·1–0·9 cm., linear or linear-lanceolate from a slightly auricled base, tapering to a very fine acute point, expanded or folded (usually towards the apex), glabrous or rarely scattered pilose, usually ciliate along the margins towards the base, finely asperulous on both surfaces. Panicles 2–9 cm. long, narrow, linear-oblong in outline, compact but sometimes interrupted in the lower part, shiny, pale green; rhachis terete, often sulcate towards the base, smooth, glabrous; branches binate or solitary, rarely in clusters, ascending to almost erect and appressed to the rhachis, glabrous, smooth. Pedicels short, with short spreading whitish hairs. Spikelets 5·5–7 mm. long (excluding the awns!), oblong-lanceolate in outline, pallid. Inferior glume c. 3·5 mm. long, 1-nerved, ovate when expanded, apex acute to acuminate, keels scabrous; superior glume 5–6·75 mm. long, 3-nerved, with the nerves close together thus leaving wide hyaline flanks, ovate when expanded, apex very acute to acuminate, keel scabrous. First and second lemmas golden yellow with silvery apices, covered with white sericeous hairs, keels scaberulous; the first c. 3·25 mm. long with the awn c. $\frac{1}{3}$–$\frac{1}{2}$ the length of the body; the second c. 2·75 mm. long, with the awn up to twice as long as the body. Anthers of the first (male) floret c. 2·8 mm. long. Terminal lemma 2–3·5 mm. long, almost invisibly 5-nerved. Anthres c. 3·25 mm. long. Caryopsis c. 1·75 mm. long.

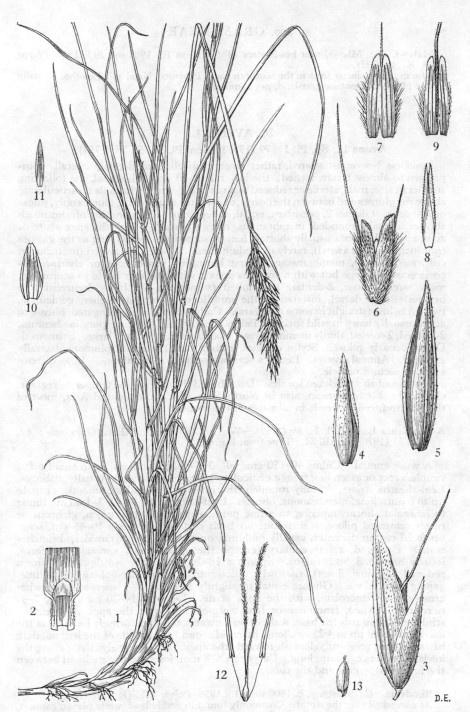

Tab. 21. ANTHOXANTHUM ECKLONII. 1, habit (×⅔); 2, junction of leaf-sheath and lamina
showing ligule (×3); 3, spikelet (×6); 4, inferior glume (×6); 5, superior glume
(×6); 6, the 2 sterile florets seen after the removal of the glumes (×6); 7, lemma of
first (♂) floret (×6); 8, palea of first floret (×6); 9, lemma of second (sterile) floret
(×6); 10, lemma of third (fertile) floret (×6); 11, palea of third floret (×6); 12,
sexual organs (×4); 13, caryopsis in lateral view (×6), all from *Robinson* 5305.

Malawi. S: Mlanje, near headwaters of Lichenya R., 1965 m., 29.iii.1960, *Phipps* 2773 (K; SRGH).

Also in mountainous areas in the southern Cape Province, Natal and Lesotho. Usually in wet places in montane grassland, not common.

20. AVENA L.

Avena L., Sp. Pl. **1**: 79 (1753); Gen. Pl. ed. 5: 34 (1754).

Spikelets 2–several-flowered, rather large, pedicelled, moderately laterally compressed to almost terete, awned; the lower 1–3(–5) florets bisexual, the following smaller in size, male, sterile or reduced to a varying degree; rhachilla disarticulating above the glumes and between the florets, or sometimes above the glumes only, pilose or glabrous. Glumes 2, persistent, equal, subequal or sometimes the inferior much shorter, dorsally rounded, membranous, margins hyaline, with the apex acute or acuminate. Lemmas usually shorter than or sometimes just as long as the glumes (not including the awns!), rarely exceeding them, 5–9-nerved, awned (muticous in cultivated forms), membraneous at first becoming chartaceous or chartaceous-coriaceous with age but with a scarious apex; with the apex acute to acuminate, more rarely obtuse, 2-dentate to 2-lobed (sometimes the lobes decurrent into bristles); awn dorsal, inserted at the middle or somewhat higher, geniculate, twisted below (straight in some cultivars). Callus short, rarely elongated, obtuse to acute, usually hairy in wild species. Paleas shorter than the corresponding lemmas, 2-nerved, 2-keeled, thinly membranous. Lodicules 2, relatively large. Stamens 3. Ovary densely pilose. Styles distinct, rather short; stigmas plumose, laterally exserted. Annual grasses. Ligule a scarious or hyaline membrane. Inflorescence an open secund panicle.

A genus of about a dozen species. Distributed throughout the temperate regions, chiefly the Mediterranean, also in North East Africa and Central Asia, most of them introduced as weeds in all parts of the globe.

Avena fatua L., Sp. Pl. **1**: 80 (1753).—W. D. Clayton in F.T.E.A., Gramineae: 82, fig. 28 (1970). TAB. **22.** Type from Europe.

A weak annual. Culms 40–150 cm. tall, 3–5-noded, solitary, or in small tufts, simple, erect or ascending from a geniculate base, terete, smooth, usually glabrous. Leaf-sheaths striate, dorsally rounded, the inferior often pilose, smooth. Ligule up to 6 mm. long, membranous, obtuse. Leaf-laminae 10–50 × 0·3–1·6 cm., linear to lanceolate-linear, tapering to a fine point, expanded, vivid green, glabrous or rarely scattered pilose, scaberulous on both surfaces. Panicle 10–45 cm. long, up to 20 cm. in diameter, usually nodding, open and loose, pyramidal; branches usually clustered, rarely solitary (except the upper ones), spreading, filiform, loosely branched, scaberulous. Spikelets 18–30 mm. long, 2–3-flowered (with a reduced terminal floret), narrowly oblong to obtriangular-oblong in outline, gaping, pendulous. Glumes equal to slightly unequal, 7–11-nerved, somewhat cross-veined, lanceolate, with the apex acute. Lemmas 14–20 mm. long, 7–9-nerved, all awned, lanceolate or lanceolate-oblong, with the apex 2–4-dentate, stiffly pilose towards the base, scaberulous towards the apex, densely bearded at the insertion; awn up to 4·25 cm. long, inserted around the middle of the lemma, dark brown below, green or yellowish towards the apex. Palea densely ciliate along the keels. Anthers c. 3 mm. long. Caryopsis 6–8 mm. long, tightly enclosed between the hardened lemma and the palea.

Rhodesia. C: Salisbury, c. 1600 m., 21.x.1958, *Phipps* 1351 (K; SRGH).

As a weed all over the world. Commonly found in arable land, waste places, gardens, along roads, and sometimes forest clearings.

In similar places *A. sativa* L. may be found. This cultivated species can be distinguished from the weedy *A. fatua* by not having the rhachilla disarticulating between the florets. Moreover the lemma is glabrous in *A. sativa*.

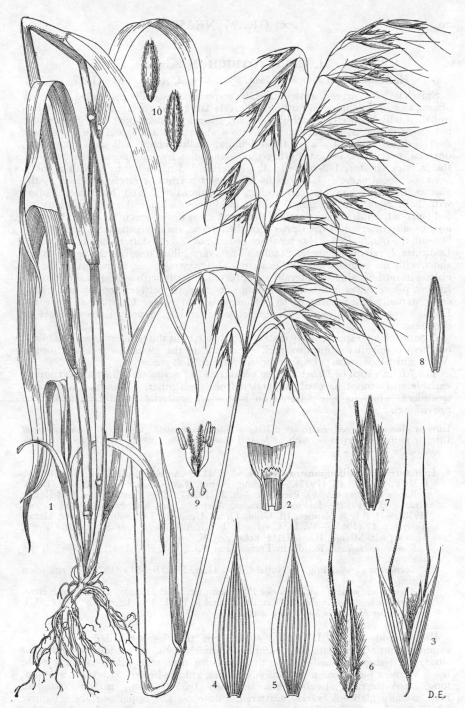

Tab. 22. AVENA FATUA. 1, habit (× ½); 2, junction of leaf-sheath and lamina showing
ligule (× 1½); 3, spikelet in lateral view (× 1½); 4, inferior glume (× 2); 5, superior
glume (× 2); 6, inferior floret (× 2); 7, lemma (× 2); 8, palea (× 2); 9, lodicules
(× 2); 10, sexual apparatus (× 2), all from *Bogdan* 1944; 11, caryopsis in dorsal and
ventral views respectively (× 2) *Phipps* 1351.

21. HELICTOTRICHON Schult.

Helictotrichon Schult., Mant. Syst. Veg. **2,** Addit.: 526 (1827).

Spikelets 2–6-flowered, awned, laterally compressed, usually pedicelled, narrowly oblong to elliptic-oblong in outline, relatively large; florets all similar, ⚥, the terminal usually reduced; rhachilla slender, disarticulating above the glumes and between the florets, pilose, extended into a short bristle beyond the uppermost floret or terminated by a vestigial lemma; callus short, villous. Glumes 2, unequal, persistent, keeled, herbaceous, sometimes hyaline, acute to acuminate; the inferior shorter, 1–3-nerved; the superior 3–5-nerved (rarely with a short, faint additional nerve). Lemmas usually exserted from or rarely enclosed by the glumes, narrowly lanceolate in profile, acute, dorsally awned, herbaceous, often with both apex and margins hyaline, rarely somewhat indurated at maturity, 5–11-nerved, 2-fid, sometimes with the apical lobes drawn out into slender bristles; awn usually inserted in the upper $\frac{1}{2}$ of the lemma, usually geniculate and twisted beneath the bend. Paleas shorter than the lemmas, 2-keeled, herbaceous to hyaline. Lodicules 2, relatively large. Stamens 3. Ovary pilose towards the apex; styles short, with the stigmas laterally exserted. Caryopsis oblong to narrowly elliptic-oblong in outline, slightly laterally compressed, usually with a longitudinal frontal furrow, pilose towards the apex, falling out enclosed between lemma and palea; embryo small; hilum linear, up to $\frac{1}{2}$ as long as the caryopsis. Caespitose perennials. Ligule a hyaline membrane. Inflorescence a panicle, usually narrow, erect or drooping.

A genus of c. 60 spp., mainly distributed throughout the temperate regions of the northern hemisphere, but also in S. Africa and in the mountain areas of tropical Africa and Madagascar; 1 sp. is recorded from Malaysia.

The African spp. of *Helictotrichon* are in need of revision. They are extremely variable and cannot be easily separated from each other. Thus any taxonomic treatment, which is solely based on herbarium material, must be regarded as provisional.

Inferior glume 1-nerved, narrowly deltate to linear-lanceolate   -   - 1. *milanjianum*
Inferior glume 3-nerved or with at least 1 lateral nerve, broadly lanceolate to ovate-
    lanceolate - - - - - - - - - - - 2. *elongatum*

1. **Helictotrichon milanjianum** (Rendle) C. E. Hubb. in Kew Bull. **1936**: 334 (1936); in F.T.A. **10**: 105 (1937).—Eggeling, Annot. List Grass. Uganda: 22 (1947). —Robyns & Tournay, Fl. Parc Nat. Alb. **3**: 182 (1955).—Bogdan, Rev. List Kenya Grass.: 30 (1958).—Jackson & Wiehe, Annot. Check-List Nyasal. Grass.: 42 (1958).—Harker & Napper, Ill. Guide Grass. Uganda: 34 (1960).—Napper, Grass. Tangan.: 17 (1965).—W. D. Clayton in F.T.E.A., Gramineae: 88 (1970). Type: Malawi, Mt. Mlanje, *Whyte* (BM, holotype; K, isotype).
        Bromus milanjianus Rendle in Trans. Linn. Soc., Bot., Ser. 2, **4**: 59 (1894). Type as above.
        Avenastrum majus Pilg. in Notizbl. Bot. Gart. Berl. **9**: 519 (1926). Type from Kenya.
        Avenastrum mannii var. *angustior* Pilg., tom. cit.: 521 (1926). Type from Kenya.
        Arrhenatherum milanjianum (Rendle) Potztal in Engl., Bot. Jahrb. **75**: 319 (1951). Type as for *Helictotrichon milanjianum.*

Rhizomatous perennial. Culms 40–125 cm. tall, 3–more-noded, erect or ascending from a geniculate base, simple or branched below, slender to somewhat stout. Leaf-sheaths usually shorter than the culm internodes, glabrous and smooth, the lowermost persistent and usually becoming golden-brown. Ligule 2·5–4 mm. long, often lacerate. Leaf-laminae 6–30(40) × 0·35–0·8 cm., usually expanded, firm, usually glabrous, rarely scattered, pilose on the upper surface. Panicle 10–27 cm. long, narrowly oblong or elliptic-oblong in outline, loose; rhachis slender, flexuous; branches filiform, usually paired, ascending to almost erect, flexuous. Spikelets 12–15 mm. long, 3–4-flowered, narrowly oblong in outline, loose, green; rhachilla internodes 2·5–3 mm. long, pilose towards the apex. Glumes not exceeding the following lemma, herbaceous with the margins thinner in texture; the inferior 3–6 mm. long, 1-nerved, narrowly lanceolate; the superior 3·75–8·25 mm. long, 3-nerved, oblong-lanceolate. Lemmas 7–12 mm. long,

7-nerved, herbaceous, linear-lanceolate in profile, with the margins hyaline; apex 2-lobed, with the lobes drawn out into slender bristles; awn inserted beneath the apex, up to 22 mm. long, somewhat geniculate, not much twisted. Paleas with the keels ciliolate. Anthers c. 2 mm. long. Caryopsis 3–4 mm. long, rather soft.

Malawi. S: Mlanje, Litchenya-Chalonwe path, 3·2 km. from Chalonwe, 1950 m., 29.iii.1960, *Phipps* 2776 (K; PRE; SRGH).

Distributed also throughout East Tropical Africa from Ethiopia to Uganda and extending into the Congo Republic. In moist places in mountain forests, damp kloofs and in bamboo thickets.

A specimen from Mlanje (Chambe), *Jackson* 1939 (K) differs in some respects, i.e. a more luxuriant growth and a more geniculate awn of the lemma, from typical *H. milanjianum* and may perhaps represent a different species. A decision on this question can only be made when more material is available.

2. **Helictotrichon elongatum** (A. Rich) C. E. Hubb. in Kew Bull. **1936**: 335 (1936); in F.T.A. **10**: 114 (1937).—Eggeling, Annot. List Grass. Uganda: 22 (1947).— Robyns & Tournay, Fl. Parc Nat. Alb. **3**: 184 (1955).—F. W. Andr., Fl. Pl. Sudan, **3**: 463 Bogdan, Rev. List Kenya Grass.: 30 (1958).—Jackson & Wiehe, Annot. Check-List Nyasal. Grass.: 42 (1958).—Harker & Napper, Ill. Guide Grass. Uganda: 34 (1960).—Napper, Grass. Tangan.: 17 (1965).—W. D. Clayton in F.T.E.A., Gramineae: 89, fig. 30 (1970). TAB. **23**. Type from Ethiopia.

    Danthonia elongata A. Rich., Tent. Fl. Abyss. **2**: 419 (1851). Type from Ethiopia.
    Trisetum neesii Steud., Syn. Pl. Glum. **1**: 227 (1854). Type from Ethiopia.
    Avena festuciformis Hochst. in Flora, **38**: 275 (1855). Type from Ethiopia.
    Avena neesii (Steud.) Hook. f. in Journ. Linn. Soc., Bot. **7**: 229 (1864). Type as for *Trisetum neesii.*
    Avena muriculata Stapf in Kew Bull. **1897**: 291 (1897) *nom. superfl.* Type from Ethiopia.
    Avenastrum elongatum (A. Rich.) Pilg. in Notizbl. Bot. Gart. Berl. **9**: 518 (1926). Type as for *Danthonia elongata.*
    Helictotrichon cartilagineum C. E. Hubb. in Kew Bull. **1936**: 331 (1936); in F.T.A. **10**: 112 (1937).—Edwards & Bogdan, Import. Grass. Pl. Kenya: 42 (1951).— Bogdan, Rev. List Kenya Grass.: 30 (1958).—Jackson & Wiehe, Annot. Check List Nyasal. Grass.: 42 (1958).—Type from Kenya.
    Helictitrichon phaneroneuron C. E. Hubb., tom. cit.: 332 (1936); in F.T.A. **10**: 109 (1937).—Napper, Grass. Tangan.: 17 (1965). Type from Tanzania.
    Arrhenatherum elongatum (A. Rich.) Potztal in Engl., Bot. Jahrb. **75**: 328 (1951).— Piovano in Webbia, **7**: 302 (1957). Type as for *Danthonia elongata.*
    Arrhenatherum phaneroneuron (C. E. Hubb.) Potztal in Wildenowia, **4**: 400 (1968) (" *phaeoneuron* "). Type as for *Helictotrichon phaneroneuron.*

A caespitose perennial. Culm 30–125(150) cm. tall, 2–5-noded, usually simple, erect or geniculately ascending, glabrous and smooth. Leaf-sheaths striate, smooth and usually glabrous, the lowermost persistent and often tinged with brown. Ligule c. 3 mm. long, rounded, entire or sometimes lacerate. Leaf-laminae 10–40 × 0·2–0·5 cm., usually expanded, tapering to a fine point, glabrous or rarely scattered pilose. Panicle 10–38 cm. long, linear, narrowly oblong to narrowly elliptic, sometimes dense and with the branches more or less appressed to the rhachis, but usually loose and open, erect or somewhat nodding; rhachis slender, flexuous; branches in groups of 2–4, filiform, scaberulous. Spikelets 8–14 mm. long, 2–3-flowered, green or purplish. Glumes firmly membranous, with the margins hyaline, acute to acuminate, usually not much shorter than the lowermost lemma; the inferior 5–10 mm. long, lanceolate-oblong, 2–3-nerved; the superior 6·5–10·5 mm. long, narrowly elliptic, 3–5-nerved. Lemmas up to 11 mm. long, chartaceous to subcartilaginous, pale green often tinged with purple, dorsally verruculose or scabrous, the 2 apical lobes either acute or drawn out into fine bristles; awn inserted dorsally around the middle, up to 20 mm. long, geniculate and twisted beneath the bend. Paleas ⅔–⅘ the length of the lemmas, narrowly oblong, with the keels ciliolate. Anthers 2·5–3·5 mm. long. Caryopsis c. 3 mm. long.

Zambia. N: Abercorn, Sunzu Mts., 32·2 km. S. of Abercorn, 1740 m., 20.iv.1961, *Phipps & Vesey-FitzGerald* 3311 (BM; K; PRE; SRGH). C: Serenje Distr., 14.v.1963, *Robinson* 5731 (K; PRE; SRGH). **Rhodesia.** C: Marandellas, Lendy Estate, 31.x.1948, *Corby* 171 (SRGH). E: Inyanga, Selborne, 1520 m., 4.i.1947, *Fisher* 1194 (PRE; SRGH). S: Bikita, Matsai Reserve, 14.iii.1956, *Cleghorn* 175 (BM; SRGH).

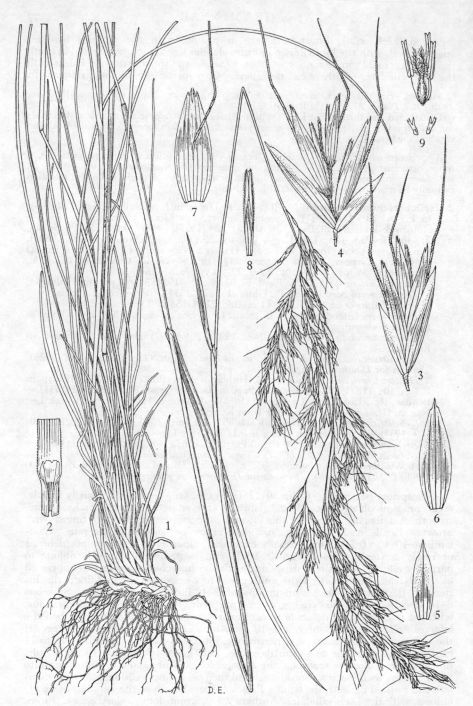

D.E.

Tab. 23. HELICTOTRICHON ELONGATUM. 1, habit (× ⅔) *Robinson* 5731 and *Siame* 53737; 2, ligule (×3); 3, spikelet (×3); 4, spikelet with florets spread to show rhachilla (×3); 5, inferior glume (×3); 6, superior glume (×3); 7, lemma (×3); 8, palea (×3); 9, sexual apparatus with lodicules detached (×3), all from *Siame* 53737.

Malawi. N: Nyika Plateau, Lake Kaulime, 2150 m., 24.x.1958, *Robson* 334 (BM; K; LISC; PRE; SRGH). S: Kirk Range, Goche, 14.xi.1950, *Jackson* 275 (K; NYAS).

Distributed also in Ethiopia, the Sudan Republic, Uganda, Kenya, Tanzania, Cameroons, the Congo Republic and Madagascar. In damp places in open woodland and mountain grassland, often along streams or in drainage channels or on river banks.

22. AIRA L.

Aira L., Sp. Pl. **1**: 63 (1753); Gen. Pl., ed. **5**: 31 (1754).

Spikelets strongly laterally compressed, pedicelled, rather small, strictly 2-flowered, awned or rarely muticous; rhachilla disarticulating above the glumes and between the florets, very shortly produced beyond the superior floret, glabrous; florets 2, both bisexual, similar. Glumes 2, subequal and similar, persistent, thinly membranous, usually hyaline, exceeding the florets except for the awns, 1-nerved, keeled, muticous. Lemmas shorter than the glumes, subequal, slightly distant, broadly lanceolate, dorsally rounded, faintly 5-nerved, very thinly membranous, hyaline, broadly lanceolate, with the apex acute to acuminate and usually shortly 2-dentate, margins usually involute, dorsally awned from beneath the middle, rarely awnless; awn geniculate, twisted below the bend (straight when moistened); callus rather short, obtuse, pilose or glabrous. Paleas slightly shorter than the corresponding lemmas, 2-nerved, 2-keeled, narrowly oblong, with the apex 2-dentate. Lodicules 2. Stamens 3. Ovary glabrous; styles distinct, short, stigmas plumose, rather short, laterally exserted. Caryopsis oblong to narrowly elliptic in outline, tightly enclosed by the lemma and palea; hilum oblong, rather small; embryo subcircular, $\frac{1}{4}$–$\frac{1}{3}$ the length of the caryopsis. Small annuals. Ligule a membrane. Inflorescence a panicle, many-flowered, usually open.

A predominantly Mediterranean genus of 8 spp., extending to northern Europe and to the mountains of tropical and S. Africa. Introduced into N. and S. America, Asia and other parts of the globe.

Aira caryophyllea L., Sp. Pl. **1**: 66 (1753).—A. Rich., Tent. Fl. Abyss. **2**: 414 (1851).—Hack. in Engl. & Prantl, Nat. Pflanzenfam. **2**, 2: 54 (1887).—Stapf in Harv. & Sond., F.C. **7**: 463 (1899).—Fiori, Nuova Fl. Anal. Ital. **1**: 103 (1923).—Pilg. in Notizbl. Bot. Gart. Berl. **9**: 515 (1926).—Peter in Fedde, Repert., Beih. **40**, 1: 302 (1931).—Maitland in Kew Bull. **1932**: 424 (1932).—Hutch. & Dalz. in F.W.T.A. **2**: 530 (1936).—C. E. Hubb. in F.T.A. **10**: 87 (1937).—Eggeling, Annot. List Grass. Uganda: 2 (1947).—Robyns & Tournay in Fl. Parc Nat. Alb. **3**: 179 (1955).—Chippindall in Meredith, Grass. & Pastures of S. Afr.: 86 (1955).—Hedberg in Symb. Bot. Upps. **15**: 46 (1957).—Bogdan, Rev. List Kenya Grass.: 31 (1958).—Jackson & Wiehe, Annot. Check List Nyasal. Grass.: 27 (1958).—Harker & Napper, Ill. Guide Grass. Uganda: 14 (1960).—Bor, Grass. of B.C.I. & P.: 430 (1960).—Napper, Grass. Tangan.: 16 (1965).—W. D. Clayton in F.T.E.A., Gramineae: 84, fig. 29 (1970). TAB. **24**. Type from Europe.

Aira latigluma Steud., Syn. Pl. Glum. **1**: 221 (1854).—Schweinf., Beitr. Fl. Aethiop.: 297 (1867). Type from Ethiopia.

Aira caryophyllea var. *latigluma* (Steud.) C. E. Hubb. in F.T.A. **10**: 88 (1937). Type as above.

A delicate loosely caespitose annual. Culms 5–30 cm. tall, 1–many-noded, rather slender, almost filiform, erect or ascending from a geniculate base, smooth, glabrous. Leaf-sheaths slightly longer or slightly shorter than the internodes, striate, tight at first later loose and slipping off the culm, scaberulous along the nerves. Ligule up to 4 mm. long, lanceolate-oblong or triangular-oblong, acute to suboblique, often lacerate. Leaf-laminae 2–7 × 0·02–0·06 cm., filiform, subsetaceous, almost always convolute, usually erect, rarely spreading, scaberulous along the nerves on both surfaces and along the margins, glabrous. Panicle 0·9–7·5 cm. long, ovate to oblong or obovate in outline, erect, somewhat contracted or open and loose; rhachis filiform, glabrous, smooth, branches paired or rarely solitary, bearing spikelets in the upper $\frac{1}{2}$, very slender, almost capillary. Pedicels 2–10 mm. long, capillary. Spikelets 2·5–4·5 mm. long, ovate to broadly oblong in lateral view, pallid green or silvery-grey, sometimes tinged with purple, slightly glossy. Glumes 2·75–4·5 mm. long, broadly lanceolate, minutely scaberulous along the keel. Lemmas 1·75–2·5 mm. long (excluding the awn), ovate-lanceolate with

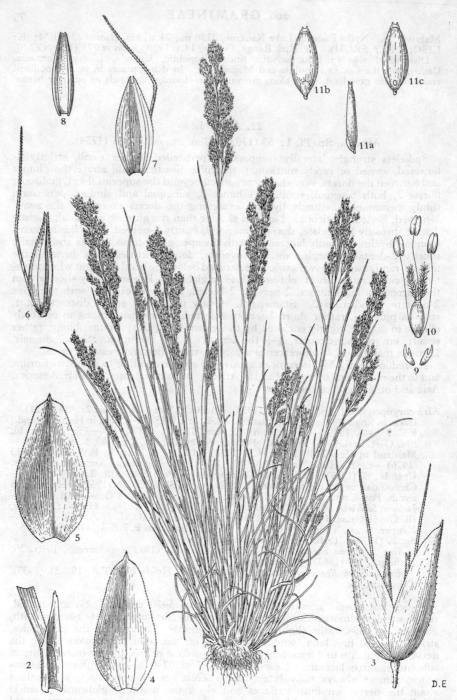

Tab. 24. AIRA CARYOPHYLLEA. 1, habit (× ½); 2, ligule (× 5); 3, spikelet (× 10); 4, inferior glume (× 10); 5, superior glume (× 10); 6, floret (× 10); 7, lemma (× 10); 8, palea (× 10); 9, lodicules (× 10); 10, sexual organs (× 10); 11, caryopsis lateral, dorsal and ventral views (× 20). 1–9 and 11 from *Jackson* 880, 10 from *Jackson* 511.

the apex acute to subobtuse, dorsally asperulous towards the apex; awn 3–4 mm. long, usually brown; callus appressed-pilose. Anthers c. 0·3 mm. long. Caryopsis c. 1 mm. long.

Malawi. N: Nyika, 28.vi.1952, *Jackson* 880 (K; SRGH).
A European species, also distributed through western Asia, northern Africa, west tropical Africa, extending throughout east tropical Africa to the Cape Prov. Along roads, in open grassland, in disturbed areas, also abandoned fields, etc., usually on thin soil.

XIII. PHALARIDEAE Kunth

Phalarideae Kunth, Rev. Gram. **1**: 12 (1829).

Inflorescence a compressed panicle, usually spike-like. Spikelets usually all similar (but sometimes arranged in dense groups in which only 1 is fertile), strongly laterally compressed, 3-flowered, with the first and second florets reduced to the lemmas, the third ♀; rhachilla disarticulating above the glumes. Glumes 2, subequal in shape and size, persistent, enclosing the rest of the spikelet. Fertile lemma strongly keeled, indurated, 5-nerved, awnless. Palea almost as long as the lemma, 2-nerved, 2-keeled, coriaceous with the margins hyaline. Lodicules 2, lanceolate. Stamens 3. Ovary glabrous; styles 2; stigmas plumose. Caryopsis loosely held within the indurated floret; embryo rather small; hilum short; starch grains compound. Chromosomes large; basic numbers 6 and 7.

The tribe consists of only 1 genus, distributed throughout warm temperate and subtropical regions of both hemispheres.

23. PHALARIS L.

Phalaris L., Sp. Pl. **1**: 54 (1753); Gen. Pl., ed. 5: 29 (1754).

Spikelets solitary, shortly pedicelled, laterally compressed, 1-flowered, but usually with 1–2 sterile and much reduced florets; rhachilla disarticulating above the glumes, not or only spuriously extended beyond the terminal floret; florets usually 3, the first and second rudimentary, minute, the terminal one ♀. Glumes 2, equal to subequal, 3-nerved, persistent, completely enclosing the floret (s), cymbiform, often strongly keeled, membranous, with the keel broadly winged. Sterile lemmas very small (often scale-like), subulate to lanceolate, membranous with a callous base, the fertile one 5-nerved, membranous at first, later coriaceous to indurate, often glossy, broadly ovate, acute. Palea c. as long as the lemma, 2-nerved, hyaline, usually ciliolate along the keels. Lodicules 2, hyaline. Stamens 3. Ovary elliptic in outline, glabrous; styles distinct, long; stigmas plumose, slender, apically exserted. Caryopsis ovate to ovate-elliptic in outline, somewhat laterally flattened, enclosed by the lemma and palea; hilum oblong or linear, short; embryo $\frac{1}{4}$–$\frac{1}{3}$ the length of the caryopsis. Annuals or perennials. Ligule membranous, usually truncate. Inflorescence a many-flowered rather condensed spike-like panicle, often capitate.

A cosmopolitan genus of 15 spp., with its evolutionary centre in the Mediterranean region.

Phalaris minor Retz., Obs. Bot. **3**: 8 (1783). TAB. **25** fig. A. Type from the Mediterranean region.
For a complete synonymy of this sp. see D. E. Anderson in Iowa State Journ. Sci. **36**: 30–31 (1961).

A loosely caespitose annual. Culms 20–100 cm. tall, 3–several-noded, rather weak, erect or geniculately ascending, striate, glabrous. Leaf-sheaths shorter than the internodes, tight at first, later loose or often somewhat inflated, striate, usually glabrous, smooth. Ligule 2–7·5 mm. long, whitish. Leaf-laminae 5–10 × 0·3–1·2 cm., linear, tapering to a fine point, flaccid, glabrous, smooth but scaberulous along the margins. Panicles 1–6 cm. long, cylindrical to subglobose, rather dense and compact; rhachis and branches glabrous. Spikelets 4–6·5 mm. long, obliquely broad-elliptic in lateral view, almost plano-convex. Glumes subequal, 4–6·5 mm.

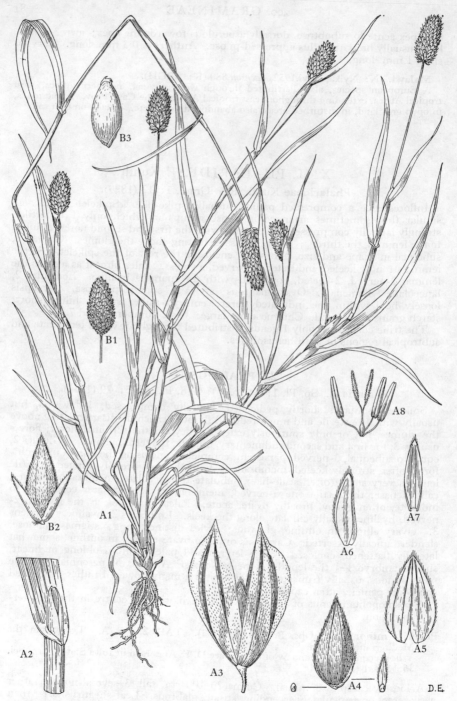

Tab. 25. A.—PHALARIS MINOR. A1, habit (× ½) *Phipps* 2218; A2, junction of leaf-sheath and lamina showing ligule (× 1½) *Hopkins* in GHS 7749; A3, spikelet (× 8); A4, floret with glumes removed to show sterile secondary floret (× 8); A5, lemma (× 8); A6, A7, palea lateral and ventral views respectively (× 8); A8, sexual organs (× 8) all from *Phipps* 2218. B.—PHALARIS CANARIENSIS. B1, inflorescence (× ½); B2, spikelet (× 5); B3, caryopsis (× 5), all from specimen cultivated at Kew.

long, with the wing erose-denticulate or sometimes entire. Fertile lemma 2·7–4·0 mm. long, dorsally pubescent; sterile floret usually 1 (the second: the first if present at all scale-like), c. 1–1·75 mm. long, subulate. Anthers 1–1·5 mm. long. Caryopsis 2·25–2·5 mm. long.

Rhodesia. N: Darwendale, Cary's Farm, 1.x.1953, *Addison* in GHS 44056 (SRGH). W: Matopos, Farm School, 1370 m., xii.1925, *Eyles* 4078 (SRGH). C: Salisbury, railway siding of S. African Timber Co., c. 1500 m., 26.ix.1959, *Phipps* 2218 (BM; SRGH). E: Penhalonga, 17.ix.1940, *Hopkins* in 7749 GHS (SRGH).

Phalaris minor, although most frequently referred to as a Mediterranean sp., is of world-wide distribution. Growing in waste places, disturbed areas, as a weed in gardens, on arable land or in pastures.

Note: The following introduced spp. also occur, often in similar localities: *P. aquatica* L. (*P. tuberosa* L.), *P. arundinacea* L. and *P. canariensis* L. They can be recognised as follows:
1 sterile floret developed (as in tab. 25, fig. A4) only:
   Plant annual; base not bulbous, wings of the glumes usually erose-denticulate   *minor*
   Plant perennial; base distinctly bulbous; wings of glumes entire   -   *aquatica*
2 sterile florets developed:
   Sterile florets unequal; the longer 1–2·2 mm. long; the shorter less than 0·5 mm. long
                                                  aquatica
   Sterile florets equal or nearly so:
      A robust perennial with creeping rhizomes. Panicle up to 20 cm. long, somewhat open, lobed or sometimes interrupted, lanceolate to oblong in outline. Sterile florets less than 1 mm. long, somewhat swollen, almost terete   -   *arundinacea*
      A loosely caespitose annual. Panicle 1·5–6 cm. long, very compact, neither lobed nor interrupted, ovate to ovate-oblong in outline. Sterile florets c. ½ as long as the fertile lemma, membranous, lanceolate - - - - - *canariensis*
P. canariensis (Tab. **25**, fig. B) is the " Canary Grass " the " seeds " of which are used as food for caged birds, a fact that accounts for the casual appearance of this sp. anywhere. *P. arundinacea* is often cultivated as a fodder grass.

XIV. AGROSTIDEAE Dumort.

Agrostideae Dumort, Obs. Gram. Belg.: 83 (1823).

Inflorescence a panicle, sometimes open but more often contracted, often spike-like. Spikelets all alike, slightly compressed laterally, 1-flowered, ☿, falling entire or disarticulating above the glumes; rhachilla sometimes produced beyond the floret simulating a slender awn. Glumes 2, usually persistent, as long as or longer than the lemma, herbaceous to membranous, 1- (rarely 3-) nerved, awnless. Lemma 3–5-nerved, membranous or often hyaline, usually with a slightly geniculate dorsal awn, sometimes awned from the apex, rarely awnless. Palea 2-nerved, 2-keeled, usually hyaline, often much smaller than the lemma or even absent. Lodicules lanceolate, 2, sometimes absent. Stamens 1–3. Stigmas 2, plumose. Caryopsis free, slightly terete, often somewhat compressed dorsally, embryo small, not more than ¼ the length of the caryopsis; hilum linear, oblong or punctiform; starch grains compound. Chromosomes large, basic number 7.

A tribe comprising c. 40 genera. Mainly distributed throughout the temperate and cold regions of both hemispheres, but also represented in mountainous areas of the tropics.

Spikelets falling entire at maturity. Glumes with very long awns or awnless.
                                                  24. **Polypogon**
Spikelets disarticulating above the glumes at maturity. Glumes awnless -   25. **Agrostis**

24. POLYPOGON Desf.

Polypogon Desf., Fl. Atlant. **1**: 66 (1798).

Spikelets pedicelled, 1-flowered, laterally compressed, gaping at length, falling entire, awned, with an obtuse to acute basal callus; rhachilla disarticulating above the glumes, not produced beyond the floret; floret ☿, shorter than the glumes

(except for the awn). Glumes 2, subequal to slightly unequal, 1-nerved, membranous, dorsally rounded below and usually keeled towards the apex, the flanks
thinner in texture than the somewhat hardened dorsal surface, with the apex
entire to shortly 2-lobed, unawned or awned from between the sinuses or slightly
below the apex if entire; awn straight, slender. Lemma awned or awnless,
shorter than the glumes (excluding the awn), obscurely 5 (very rarely 3-)-nerved,
thinly membranous, hyaline, with the apex truncate, the lateral nerves usually
excurrent into short mucros; awn straight, slender, usually deciduous, rarely
reduced to a mucro or completely absent. Palea as long as or slightly shorter than
the body of the lemma, 2-nerved, 2-keeled, very thinly membranous, hyaline.
Lodicules 2, rather small. Stamens 3. Ovary glabrous; styles distinct, very short;
stigmas short, plumose, laterally exserted. Caryopsis elliptic to obovate-elliptic in
outline, terete in cross-section, free between the slightly indurate lemma and palea;
hilum short; embryo $\frac{1}{4}$–$\frac{1}{3}$ the length of the caryopsis. Annuals or perennials with
slender culms. Ligule a scarious membrane. Inflorescence a many-flowered
panicle, usually condensed, rarely somewhat open.

A genus of c. 10 spp., distributed throughout the temperate and warm regions of
the world.

Glumes distinctly awned; the awns several times longer than the glumes. Panicle densely
    contracted, bottle-brush-like, rarely interrupted. Plant annual -   1. *monspeliensis*
Glumes not awned or rarely with short awn-like tips. Panicle less contracted, often
    interrupted or lobed. Plant perennial - - - - - 2. *semiverticillatus*

1. **Polypogon monspeliensis** (L.) Desf., Fl. Atlant. **1**: 67 (1798).—Hack. in Engl. &
Prantl, Nat. Pflanzenfam. **2**, 2: 50 (1881).—Rendle in Cat. Afr. Pl. Welw. **2**, 1:
206 (1899).—Stapf. in Harv. & Sond., F.C. **7**: 543 (1899).—C. E. Hubb. in F.T.A.
10: 160 (1937).—Chippindall in Meredith, Grasses & Pastures of S. Afr.: 102, 103
(1955).—F. W. Andr., Fl. Pl. Sudan, **3**: 521 (1956).—Bogdan, Rev. List Kenya Grass.:
31 (1958).—Bor, Grass. B.C.I. & P.: 403 (1960).—Napper, Grass. Tangan.: 19
(1965).—C. E. Hubb. in F.T.E.A., Gramineae: 100 (1970).—Launert in Merxm.,
Prodr. Fl. S. W. Afr. **160**: 154 (1970). TAB. **26**. Type from France.
    Alopecurus monspeliensis L., Sp. Pl. **1**: 61 (1753). Type as above.
    Phleum crinitum Schreb., Beschr. Graes. **1**: 151 (1769) *nom. illegit.*
    Phalaris cristata Forsk., Fl. Aegypt.-Arab.: LX, 17 (1775). Type from Egypt.
    Agrostis alopecuroides Lam., Tabl. Encycl. Méth., Bot. **1**: 160 (1791). Type from
France or southern Europe.
    Santia plumosa Savi in Mem. Soc. Ital. Sci. **8**: 479 (1799). Type from Italy.
    Phleum monspeliense (L.) Koel., Descr. Gram.: 57 (1802). Type as for *Polypogon
monspeliensis.*
    Polypogon monspeliensis var. *capensis* Steud. in Flora, **12**: 466 (1829). Type from
S. Africa.
    Polypogon polysetus Steud., loc. cit. Type from Europe.
    Polypogon flavescens J. S. Presl ex C. B. Presl, Reliq. Haenk. **1**: 234 (1830). Type
from Peru.
    Santia monspeliensis (L.) Parl., Fl. Palerm. **1**: 73 (1845).—Type as for *Polypogon
monspeliensis.*
    Polypogon monspeliensis var. *vulgaris* Coss. in Coss. & Dur., Expl. Sci. Algér.
Glumac.: 69 (1855). Type from southern ? Europe.

A loosely caespitose rather variable annual. Culms 5–75(90) cm. tall, 2–6-noded,
erect or ascending from a decumbent base, slender to somewhat stout, simple,
glabrous, smooth or sometimes scaberulous towards the panicle. Leaf-sheaths
longer (the lower) to shorter (the upper) than the internodes, usually loose, glabrous,
smooth or scaberulous towards the mouth. Ligule 3–15 mm. long, oblong, often
lacerate, dorsally asperulous. Leaf-laminae 2·5–15(20) × 0·2–0·9 cm., lanceolate-
linear to linear, tapering to an acute point, usually expanded, flaccid to somewhat
rigid, scaberulous along the nerves on both surfaces or smooth on the lower
surface. Panicle 1·5–16 cm. long, very dense, spike-like, bristly, cylindrical or
often lobed, compact, rarely interrupted, pale-green or yellowish, rarely brownish;
branches obliquely ascending to almost appressed to the rhachis, divided from the
base. Lateral pedicels rather short, disarticulating about the middle. Spikelets
2–3 mm. long (excluding the awn). Glumes 1–2 mm. long (excluding the awn),
subequal, narrowly oblong, dorsally scabrous, with the obtuse apex minutely
2-lobed; awn 4–7·5 mm. long, scaberulous. Lemma c. $\frac{1}{2}$ the length of the glumes

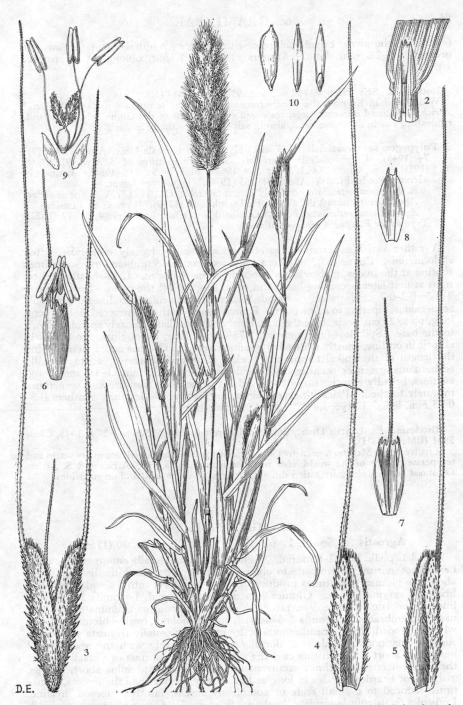

D.E.

Tab. 26. POLYPOGON MONSPELIENSIS. 1, habit (×½); 2, junction of leaf-sheath and lamina, showing ligule (×5), from *Bally* 5547; 3, spikelet in lateral view (×20); 4, inferior glume (×20); 5, superior glume (×20); 6, floret (×20); 7, lemma (×20); 8, palea (×20); 9, sexual organs with lodicules detached (×20) all from *Bally* 5547; 10, caryopsis, lateral, ventral and dorsal views respectively (×20) from *Welwitsch* 2620.

(excluding the awn), broadly elliptic with the apex minutely dentate; awn (if present) 1·5–2·5 mm. long. Anthers c. 0·5 mm. long, oblong. Caryopsis c. 1 mm. long.

Botswana. SE: Palapye, 760 m., 6.ix.1954, *Story* 4615 (K).
Widespread in Europe, the Mediterranean and in the temperate parts of Asia and Africa, introduced into most temperate and warm regions of the Globe. In waste land, rubbish tips or in cultivated land, also in salt marshes; usually in damp soil.

2. **Polypogon semiverticillatus** (Forsk.) Hylander in Uppsala Univ. Årsskr. **1945** (7): 74 (1945).—Chippindall in Meredith, Grasses & Pastures of S. Afr.: 97, 98, 99 (1955).—Bor, Grass. B.C.I. & P.: 389 (1960); Fl. Iraq, **9**: 318 (1968).—Launert in Merxm., Prodr. Fl. SW. Afr. **160**: 155 (1970). Type from Egypt.
     Phalaris semiverticillata Forsk., Fl. Aegypt.-Arab.: LX, 17 (1775). Type as above.
     Agrostis verticillata Vill., Prosp. Pl. Dauph.: 16 (1779). Type from France.
     Agrostis semiverticillata (Forsk.) C. Christ. in Dansk Bot. Ark. **4**, 3: 12 (1922). Type as for *Polypogon semiverticillatus*.

A rather variable perennial, rarely annual, usually loosely caespitose, often stoloniferous. Culms (15)30–100 cm. tall, erect from a decumbent base, sometimes rooting at the nodes, glabrous and smooth, 2–many-noded. Leaf-sheaths striate, tight at first later becoming loose and often slipping off the culm. Leaf-laminae 2–17 × 0·2–0·7 cm., usually expanded narrowly lanceolate to linear, glabrous, scaberulous, tapering to a fine point. Panicle usually rather contracted, 3–8(11) cm. long, up to 25 cm. wide, lobed, erect; branches verticillate, densely spiculate down to the base. Spikelets falling entire, 1·75–2 mm. long, ovate-elliptic to narrowly elliptic in outline, shortly pedicelled to almost sessile. Glumes equal, determining the length of the spikelet, obtusely keeled, elliptic if expanded, acute, dorsally scaberulous, green or occasionally tinged with purple. Lemma 1–1·5 mm. long, awnless, broadly elliptic, finely 5-nerved, with the apex truncate and sometimes minutely toothed. Palea 2-nerved, almost as long as the lemma. Anthers 0·5– 0·75 mm. long. Caryopsis c. 1 mm. long, pale-brown.

Rhodesia. E: Umtali Distr., Stapleford Forest Reserve, 1520 m., 29.ix.1948, *Chase* 5381 (BM; SRGH).
A native of the Mediterranean area and NE. Africa but introduced into many warm and temperate regions of the world, also recorded from Angola, SW. Africa and S. Africa. On moist ground, along irrigation ditches, streams, in waste land and on roadsides.

25. AGROSTIS L.
Agrostis L., Sp. Pl. **1**: 61 (1753); Gen. Pl. ed. 5: 30 (1754).

Spikelets pedicelled, 1-flowered, ☿, relatively small, laterally compressed, awned or awnless, narrowly lanceolate to oblong, often gaping; rhachilla disarticulating above the glumes, sometimes produced beyond the floret into a slender awn-like bristle of varying length. Glumes 2, persistent, subequal, 1- (rarely 3-) nerved, linear-lanceolate to lanceolate, rarely oblong, apex acute or acuminate, awnless, usually membranous. Lemma 3–5-nerved, broadly oblong, ovate-oblong or elliptic when expanded, thinly membranous or hyaline, apex usually truncate sometimes slightly emarginate, awnless or dorsally awned or rarely with the apex awned between 2 short lobes (in this case the awn straight), glabrous or dorsally pilose, the lateral nerves sometimes excurrent into short mucros; callus short, obtuse, glabrous or bearded. Palea as long as but usually shorter than the lemma, sometimes reduced to a small scale or completely absent, faintly 2-nerved, hyaline. Lodicules 2, usually lanceolate, hyaline. Stamens 3. Ovary glabrous; styles 2, rather short, with the stigmas laterally exserted. Caryopsis free, oblong or elliptic-oblong in outline, longitudinally grooved; embryo short, not longer than ¼ the length of the caryopsis; hilum basal, punctiform or linear to narrowly oblong. Annual or perennial grasses. Ligule a membrane, sometimes lacerate, scarious to hyaline. Inflorescence a many-flowered panicle, usually much divided, effuse or contracted, seldom spike-like.

A genus of more than 150 spp.; distributed throughout the temperate and cold regions of the world but mainly in the northern hemisphere, extending with a few spp. into the higher altitudes of the tropics.*

Lemma awnless or rarely with 1–5 rather short mucros up to 0·75 (rarely 1·5) mm. in length:
    Lemma 3-nerved, pilose or glabrous. Callus bearded - - - 1. *lachnantha*
    Lemma 5-nerved, glabrous. Callus glabrous - - - - - 2. *schimperana*
Lemma always distinctly awned:
    Panicle contracted, almost spike-like. Rhachilla not or only shortly produced beyond the floret. Lemma glabrous or rarely scattered pilose. Palea ¼–½ (sometimes ⅔) the length of the lemma - - - - - - - 3. *continuata*
    Panicle effuse, rarely somewhat contracted but never spike-like. Rhachilla usually clearly produced beyond the floret, forming an awn-like bristle (1) 2–3 mm. long; rarely the rhachilla not produced. Palea as long as or only slightly shorter than the lemma, very rarely ¼–½ the length of the lemma - - - - 4. *producta*

1. **Agrostis lachnantha** Nees in Linnaea, **10**, Litt.: 115 (1836).—Dur. & Schinz, Consp. Fl. Afr. **5**: 828 (1895).—Stapf in Harv. & Sond., F.C. **7**: 549 (1899).—Stent in Bothalia, **1**, 4: 283 (1924).—C. E. Hubb. in F.T.A. **10**: 172 (1937).—Eggeling, Annot. List Grass. Uganda: 1 (1947).—Sturgeon in Rhod. Agric. Journ. **51**: 224 (1954).—Robyns & Tournay in Fl. Parc Nat. Alb. **3**: 176 (1955).—Chippindall in Meredith, Grasses & Pastures of S. Afr.: 101 (1955).—F. W. Andr., Fl. Pl. Sudan, **3**: 387 (1956).—Jackson & Wiehe, Annot. Check List Nyasal. Grass.: 27 (1958).—Harker & Napper, Ill. Guide Grass. Uganda: 13 (1960).—Napper, Grass. Tangan.: 18 (1965).—W. D. Clayton, F.T.E.A., Gramineae: 106 (1970).—Launert in Merxm., Prodr. Fl. SW. Afr. **160**: 20 (1970). Type from S. Africa (Cape Prov.)
    Agrostis vestita A. Rich., Tent. Fl. Abyss. **2**: 401 (1851). Type from Ethiopia.
    Calamagrostis welwitschii Rendle, Cat. Afr. Pl. Welw. **2**: 205 (1899), non *Agrostis welwitschii* Steud. (1854). Type from Angola.
    Calamagrostis huttoniae Hack. in Rec. Albany Mus. **1**: 340 (1905). Type from S. Africa (Natal).
    Agrostis papposa Mez in Fedde, Repert. **18**: 2 (1922). Syntypes from S. Africa.
    Agrostis huttoniae (Hack.) C. E. Hubb. ex Goossens & Papendorf in S. Afr. Journ. Sci. **41**: 179 (1945).—C. E. Hubb. in F.T.A. **10**: 172 (1937) in clavi.—Sturgeon in Rhod. Agr. Journ. **51**: 224 (1954).—Chippindall in Meredith, Grasses & Pastures of S. Afr.: 100, 101 (1955).—Jackson & Wiehe, Annot. Check List Nyasal. Grass.: 27 (1958).—Type as for *Calamagrostis huttoniae*.

A loosely caespitose perennial or annual. Culms 25–90(110) cm. tall, erect or ascending, often rooting at the lower nodes, 2–several-noded. Leaf-sheaths tight at first, later becoming loose and sometimes slipping off the culm. Leaf-laminae 6–25 × 0·2–0·5 cm., usually expanded, but sometimes tightly folded, flaccid, pale green. Panicle 5–35 × 1–3 cm., slightly drooping, somewhat contracted, narrowly oblong to elliptic-oblong in outline. Spikelets 2–2·5 mm. long, pale-green or silvery-green sometimes tinged with purple, somewhat glossy, gaping. Glumes persistent, determining the length of the spikelet, narrowly lanceolate-acute, strongly keeled, flanks smooth or slightly granular, keels scabrid. Lemma 1·5–2·25 mm. long, 3-nerved, awnless or with the central nerve excurrent into a short subapical bristle up to 0·5 mm. long, ovate-oblong when expanded, dorsally pilose or pubescent, rarely glabrous, apex denticulate; callus bearded with hairs of varying length. Palea not much shorter than the lemma. Anthers 0·5–0·8 mm. long, oblong.

Zambia. C: Lusaka, Chalimbana, riverside rocks, 1300 m., 19.v.1957, *Robinson* 2210 (PRE; SRGH). **Rhodesia.** N: Mazoe, Makhalanga Farm, xi.1958, *Pollitt* in GHS. 91015 (BM; PRE; SRGH). W: Matobo, 1400 m., xi.1957, *Miller* 4801 (COI; LISC; SRGH). C: Enterprise, in damp crevices of granite bed of Umwindsi R., 18.viii.1946, *Wild* 1193 (K; SRGH). E: Melsetter, 1830 m., 27.xii.1948, *Chase* 1142 (COI; K; LISC; SRGH). **Malawi.** N: near Nganda Hill, Nyika Plateau, close to stream in valley 1·6 km. N. of track between Lake Kaulime and Nganda Hill, 19·3 km. from Lake

* In complete agreement with Dr. Clayton (F.T.E.A., Gramineae, p. 104 (1970)) it should be stated that our knowledge of the African spp. of *Agrostis* is still rather scanty. Both the nature and the range of variability is incompletely known, and many spp. are more variable than was assumed by previous workers. Only a taxonomic revision of the genus in Africa in conjunction with both field-observations and cytological investigations will elucidate the remaining problems.

Kaulime, 2320 m., 7.ix.1962, *Tyrer* 881 (COI; SRGH). S: Kirk Range, Gochi, 14.xi. 1950, *Jackson* 699 (SRGH).

Also known from Ethiopia, the Sudan Republic, Uganda, Tanzania, SW. Africa and S. Africa. In water or in wet soil along or in streams, on riverbanks, near waterfalls, in irrigation canals, on wet rocks, in kloof forest and along roads.

A. huttoniae is described as differing from *A. lachnantha* in having longer callus-hairs and the rhachilla produced beyond the floret. I have found so many intermediate stages here that I am quite unable to separate the 2 taxa.

2. **Agrostis schimperana** Steud., Syn. Pl. Glum. **1**: 170 (1854).—Engl., Pflanzenw. Ost-Afr. **A**: 100 (1895).—C. E. Hubb. in F.T.A. **10**: 175 (1932).—Eggeling, Annot. List Grass. Uganda: 1 (1947).—Robyns & Tournay, Fl. Parc Nat. Alb. **3**: 172 (1955).—Hedberg, Afroalpine Vasc. Pl.: 47 (1957).—Bogdan, Rev. List Kenya Grass.: 32 (1958).—Harker & Napper, Ill. Guide Grass. Tangan.: 13 (1960).— Napper, Grass. Tangan.: 18 (1965).—W. D. Clayton, F.T.E.A., Gramineae: 105 (1970). Type from Ethiopia.
    Agrostis simensis Steud., tom. cit.: 173 (1854). Type from Ethiopia.
    Agrostis hirtella Steud., loc. cit. Type from Ethiopia.
    Agrostis alba var. *schimperana* (Steud.) Engl., Hochgebirgsfl. Trop. Afr.: 128 (1892). Type as for *Agrostis schimperana*.
    Agrostis alba var. *simensis* (Steud.) Engl., loc. cit. Type as for *Agrostis simensis*.
    Agrostis schimperana var. *carinata* Engl., in Abh. Preuss. Akad. Wiss. **1894**: 58 (1894) *nom. nud.*
    Agrostis stolonifera var. *schimperana* (Steud.) Peter, Fl. Deutsch Ost-Afr. **1**: 295 (1931) *in syn.*

A stoloniferous perennial, loosely caespitose. Culms 25–140 cm. tall or even more, erect or geniculately ascending, slender, simple, 3–many-noded. Leaf-sheaths scabridulous towards the mouth. Leaf-laminae 5–17 × 0·25–0·6 cm., usually expanded, tapering to a fine point, glabrous, sometimes slightly rigid, scaberulous. Panicle 5–25 cm. long, up to 2·75 mm. wide, contracted, linear to oblong in outline, straight or slightly curved, sometimes lobed or interrupted. Spikelets 2–2·6 mm. long, widely gaping, pale-green or tinged with purple. Glumes 1-nerved, subequal, as long as the spikelet, lanceolate, membranous, acute, scaberulous to scabrous. Lemma 1·4–1·6 mm. long, oblong-elliptic when expanded, 5-nerved, apex truncate, all nerves or at least the lateral pairs excurrent into short mucros, rarely with the central nerve excurrent into a short (0·5 mm.) awn; callus glabrous. Palea slightly shorter than the lemma, faintly 2-nerved, apex usually 2-dentate; rhachilla not produced beyond the floret. Anthers 0·8–1·4 mm. long.

Zambia. C: Serenje, Kundalila Falls, 13.x.1963, *Robinson* 5696 (K; PRE: SRGH), E: Nyika Plateau, 2130 m., 2.i.1959, *Robinson* 2998 (BM; K; SRGH). S: Muckle Neuk. 19·3 km. N. of Choma, 1280 m., 27.xi.1954, *Robinson* 974 (SRGH). **Rhodesia.** E: Melsetter, Fairview Farm, c. 1920 m., 15.xi.1950, *Crook* 285 (SRGH).

Also known from Ethiopia, Uganda, Kenya and the mainland of Tanzania. In moist places in mountain grassland, along streams and on river banks, on wet rocks near waterfalls and in gorges.

3. **Agrostis continuata** Stapf in Kew Bull. **1897**: 290 (1897).—C. E. Hubb. in F.T.A. **10**: 180 (1937).—Sturgeon in Rhod. Agr. Journ. **51**: 224 (1954).—Jackson & Wiehe, Annot. Check List Nyasal. Grass.: 27 (1958).—W. D. Clayton, F.T.E.A., Gramineae: 111 (1970). Type: Malawi; *Buchanan* 356 (K, holotype).
    Agrostis natalensis Stapf in Kew Bull. **1897**: 290 (1897). Type from S. Africa (Natal).
    Agrostis radula Mez in Fedde, Repert. **18**: 2 (1922). Type: Malawi, without precise locality, *Buchanan* (B, holotype).
    Agrostis makoniensis Stent & Rattray in Proc. Rhod. Sci. Ass. **32**: 43 (1933). Type: Rhodesia, Umtali Distr., Stapleford, *Brain* 9763 (K, klastotype; SRGH).
    Agrostis whytei C. E. Hubb. in Kew Bull. **1936**: 302 (1936).—Jackson & Wiehe, Annot. Check List Nyasal. Grass.: 27 (1958). Type: Malawi, Mt. Zomba, 1300–2000 m., *Whyte* (K, holotype).

A caespitose perennial. Culms 25–60(80) cm. tall, erect or ascending from a slightly geniculate base, rather slender, simple or rarely branched, 2–4-noded. Leaf-sheaths fairly tight, smooth or scaberulous especially towards the mouth. Leaf-laminae 7–25 × 0·2–0·3 cm., usually expanded, tapering to a somewhat pungent point, firm.   Panicle 5–22 cm. long, erect, straight, contracted, spike-like, with the branches densely fascicled and densely spiculate.   Spikelets 3–5 mm. long, green, often tinged with purple, gaping.   Glumes 1-nerved, slightly unequal in size (the inferior slightly longer), lanceolate, apex acute, flanks scaberulous, scabrous along the keel.   Lemma 2–2·5 mm. long, 5-nerved, ovate-oblong, thinly membranous, apex with the lateral nerves excurrent into short mucros, glabrous or somewhat pilose, dorsally awned from the lower ½; awn 2–4·5 mm. long, geniculate. Palea ¼–½ (rarely ⅘) the length of the lemma, faintly 2-nerved, with the apex 2-dentate.   Anthers 1–1·5 mm. long.

Rhodesia. C: Marandellas, Digglefold, 5.i.1949, *Corby* 336 (K; SRGH). E: Inyanga Distr., Mare R., 1710 m., 17.i.1951, *Chase* 3630 (BM; SRGH). **Malawi.** 1891, *Buchanan* 356 (K).
Also known from the mainland of Tanzania.   In wet vleis, in swamps, on streamsides and in similar damp localities.

The plant described as *Agrostis whytei* is still only represented by the type specimen. It differs from the rest of the material by having a less contracted panicle and the lemma with the awn inserted about half-way up.   Except for these really minor differences the specimen cannot be distinguished from *A. continuata*.

4. **Agrostis producta** Pilg. in Engl., Bot. Jahrb. **39**: 600 (1907).—R. E. Fr., Wiss. Ergebn. Schwed. Rhod.-Kong. Exped. **1**: 208 (1916).—C. E. Hubbard in F.T.A. **10**: 189 (1937).—Edwards & Bogdan, Imp. Grass. Pl. Kenya: 44 (1951).—F. W. Andr., Fl. Pl. Sudan, **3**: 386 (1956).—Bogdan, Rev. List Kenya Grass.: 32 (1958).— Jackson & Wiehe, Annot. Check-List Nyasal. Grass.: 27 (1958).—Harker & Napper, Illust. Guide Grass. Uganda: 14 (1960).—Napper, Grass. Tangan.: 18 (1965).— W. D. Clayton in F.T.E.A., Gramineae: 108 (1970). TAB. **27**. Syntypes from Kenya and Tanzania.
Agrostis greenwayi C. E. Hubb. in Kew Bull. **1949**: 342 (1949).—Jackson & Wiehe, Annot. Check-List Nyasal. Grass.: 27 (1958). Type: Malawi, Mt. Mlanje, Lichenya Plateau, 16.x.1941, *Greenway* 6307 (K, holotype).

A densely caespitose perennial.   Culms 50–120 cm. tall, usually erect, more rarely ascending from a shortly geniculate base, 2–4-noded, usually simple, smooth or scaberulous beneath the nodes.   Leaf-sheaths fairly tight, smooth or scaberulous, the oldest persistent and often splitting into irregular fibres.   Leaf-laminae 3–25 × 0·25(0·3) cm., expanded or sometimes involute, linear to almost setaceous, glabrous.   Panicle 5–25 cm. long, ovate to elliptic in outline, usually effuse with flexuous branches, more rarely slightly contracted.   Spikelets 3·5–4·5 mm. long, oblong, green or often purplish, gaping at length.   Glumes 1-nerved (rarely the superior 3-nerved), subequal, lanceolate, firmly membranous, apex acute to acuminate, keel scabrous, flanks minutely granular-scaberulous or very rarely smooth. Lemma 2–2·25(3) mm. long, thinly membranous, ovate-oblong to broadly elliptic, 5-nerved with the nerves excurrent into mucros up to 0·75 mm. long, dorsally pilose to a varying degree of density, dorsally awned from near the base; awn geniculate, 4–7·5 mm. long.   Palea usually almost as long as (rarely shorter than— see notes) the lemma, faintly 2-nerved, apex 2-dentate; rhachilla usually produced beyond the floret (very rarely not produced), forming a basally hairy awn-like bristle, 1–3 mm. long.

Rhodesia. E: Manicaland, *Gilliland* 877 (K). **Malawi.** N: Rumpi, Lake Kaulime, Nyika Plateau, 4.i.1959, *Robinson* 3040 (SRGH). S: Mlanje, 2·5 km. W. of Lichenya Forestry Hut, 1800 m., 28.iii.1960, *Phipps* 2767 (K; PRE; SRGH).
Distributed throughout East Africa from the Sudan Republic, Uganda, Kenya, the mainland of Tanzania to Malawi and Rhodesia.   In montane grassland, on hillsides and rocky slopes, often to be found in shade under bracken, also along roads.

As pointed out by W. D. Clayton (F.T.E.A., Gramineae: 110), *A. producta* is a rather variable plant and not always easy to separate from other spp., notably from *A. kilimand-scharica*.   The specimen described by C. E. Hubbard as *A. greenwayi* has reluctantly been put into the synonymy of *A. producta*; although it has virtually no rhachilla extension beyond the floret I could not find any other features that would justify its specific separation from this species.   The fact that the palea in *A. greenwayi* is shorter than is usual in *A. producta* is certainly not of taxonomic importance.

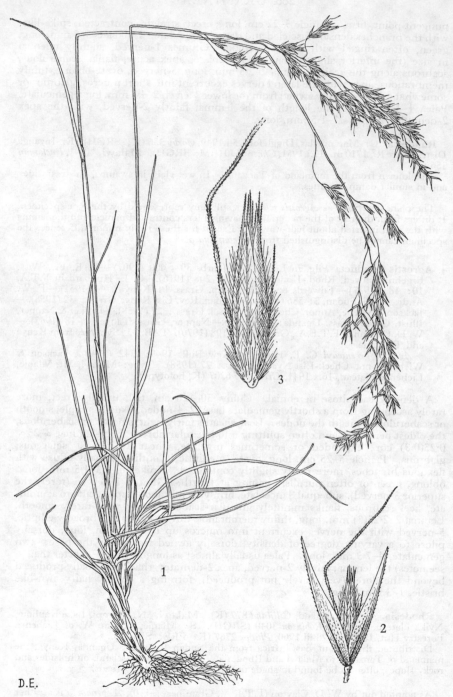

Tab. 27. AGROSTIS PRODUCTA. 1, habit (× ⅔); 2, spikelet (× 6); 3, floret, showing rhachilla-extension (× 12) all from *Bogdan* 4977. From F.T.E.A.

XV. ARUNDINEAE Dumort.

Arundineae Dumort., Obs. Gram. Belg.: 82, 124 (1823).

Inflorescence large plumose panicles. Spikelets 2–10-flowered, slightly laterally compressed, ⚥ or unisexual and dioecious, with the florets all alike or occasionally the lowermost sterile or ♂, often penicillate with long sericeous hairs; rhachilla disarticulating below each floret. Glumes 2, persistent, usually shorter than the spikelets, membranous or hyaline, similar in shape and size or the inferior smaller. Lemmas similar to the glumes, 1–5 (rarely more-) nerved, dorsally glabrous or villose, awnless or awned from the apex, often acuminate. Paleas much shorter than the corresponding lemmas, 2-keeled. Lodicules 2 or 3, obovate. Stamens 2 or 3. Ovary glabrous; styles 2; stigmas plumose. Caryopsis loose within the floret; embryo large; hilum short, oblong, basal. Starch grains compound. Chromosomes small; basic number 12.

A taxonomically not at all homogeneous tribe of c. 5 genera. Its delimitation from the *Danthonieae* is rather obscure. Distributed throughout temperate and tropical regions of both hemispheres.

26. PHRAGMITES Adans.

Phragmites Adans., Fam. Pl. **2**: 34, 559 (1763).

Spikelets pedicelled, solitary, 3–11-flowered, laterally compressed, muticous, gaping at maturity; rhachilla disarticulating above the glumes and between the florets; inferior floret sterile or ♂, the following bisexual but the uppermost usually reduced. Glumes 3–5-nerved, sometimes with cross-veins, subequal to unequal, persistent, membranous, lanceolate, obovate to oblong, dorsally rounded, apex acute, subacute or sometimes obliquely truncate. Lemmas heteromorphous, dorsally rounded, always glabrous, membranous; the inferior much longer than the inferior glume, usually persistent, 3 (sometimes 5–7-) nerved, linear-lanceolate, lanceolate-oblong or oblong, rather acute; the following (fertile), caducous, 3–1-nerved, linear lanceolate, with the apex acuminate or caudate-acuminate, thinly membranous; callus slender, straight or slightly curved, obtuse, densely bearded with long soft silky hairs. Paleas much shorter than the corresponding lemmas, 2-keeled, oblong, hyaline. Lodicules 2, glabrous. Stamens 3 (2 in the inferior floret if present). Ovary glabrous; styles distinct, rather short; stigmas plumose, laterally exserted. Caryopsis oblong in outline, semiterete, loosely enclosed between the lemma and palea; hilum basal, oblong, small; embryo about ½ as long as the caryopsis. Stout aquatic or semi-aquatic perennials with creeping much-branched rhizomes. Culms rather tall, robust, sheathed. Ligule a short membrane, densely ciliate along the apex. Inflorescence a dense or loose many-flowered panicle.

A genus of 3 spp. distributed throughout the tropical and temperate regions of both hemispheres.

All species of this genus display a considerable variability. Because of a remarkable overlap none of the characters, which are set out in the following key, can always be used reliably for the identification of the species, but in correlation with the others one always arrives at satisfactory results. It is quite possible that near the boundary of each species introgression may take place (see also W. D. Clayton in Kew Bull. **21**: 113 (1967)).

Hairs of the callus 6–10 mm. long, rather dense. Superior glume 5–9 mm. long, very
  different from the inferior. Leaf-laminae usually smooth on the lower surface, apex
  filiform, flexuous -   -   -   -   -   -   -   -   -   1. *australis*
Hairs of callus 4–7 mm. long, less dense. Superior glume 3–5 mm. long, equal or subequal
  to the inferior. Leaf-laminae scabrous on the lower surface (at least towards the
  apex), apex attenuate, somewhat stiff to almost pungent -   -   - 2. *mauritianus*

1. **Phragmites australis** (Cav.) Trin. ex Steud., Nom. Bot., ed. 2, **2**: 324 (1841).—
   Greuter & Rechinger in Boissiera, **13**: 174 (1967).—Launert in Merxm., Prodr. Fl.

SW. Afr. **160**: 151 (1970).—W. D. Clayton, F.T.E.A., Gramineae: 116 (1970). Type from South Australia.

Arundo phragmites L., Sp. Pl. **1**: 81 (1753). Type from Europe.

Arundo vulgaris Lam., Fl. Franç. **3**: 615 (1778). *nom. superfl.* (based on *A. phragmites*). Type from Europe.

Arundo australis Cav. in Anal. Hist. Nat. Madrid, **1**: 100 (1799). Type as for *Phragmites australis*.

Phragmites communis Trin., Fund. Agrost.: 134 (1820).—Hack. in Engl. & Prantl, Nat. Pflanzenfam. **2**, 2: 68 (1887).—Stapf in Harv. & Sond., F.C. **7**: 541 (1899).—Bremekamp & Oberm. in Ann. Transv. Mus. **16**: 405 (1935).—C. E. Hubb. in F.T.A. **10**: 153 (1937).—Sturgeon in Rhod. Agric. Journ. **51**: 215 (1954).—Chippendall in Meredith, Grasses & Pastures of S. Afr.: 228 (1955).—Bor, Grasses of B.C.I. & P.: 416 (1960).—Conert in Syst. Anat. Arund.: 40 (1961). Type from Europe.

Phragmites vulgaris (Lam.) Crép., Man. Fl. Belg., ed. 2: 345 (1866).—R. E. Fr. in Wiss. Ergebn. Schwed. Rhod.-Kong.-Exped. **1**: 210 (1916).—Hutch. & Dalz., F.W.T.A. **2**: 510 (1931). Type as for *Arundo vulgaris*.

Trichoon phragmites (L.) Rendle, Cat. Afr. Pl. Welw. **2**: 218 (1899). Type as for *Arundo phragmites*.

For a comprehensive synonomy see Conert (loc. cit.), and Greuter & Rechinger (loc. cit.).

A stout perennial growing with a system of long creeping scaly rhizomes. Culms (0·3)0·6–6 m. tall or more, up to 1·5 cm. in diam., many-noded, slender to stout, usually erect but sometimes floating, simple, terete, glabrous. Leaf-sheaths imbricate at first, later shorter than the internodes, striate, tight, smooth, glabrous. Ligule c. 1 mm. long. Leaf-laminae 4–100 × 0·3–5 cm., linear to very narrowly lanceolate, slightly auriculate at the base, tapering to a long filiform flexuous point, usually expanded, rather firm, glabrous but sometime pilose near the base, smooth on both surfaces but sometimes scaberulous along the edges. Panicle 10–50 × 3·5–17 cm., silvery-purplish or brownish, oblong to ovate-oblong in outline, erect or nodding later on, rather dense; rhachis terete in the lower, angular in the upper part, barbate at the nodes; branches fascicled, repeatedly divided, angular, some of them bearing spikelets almost to their base. Pedicels 1–8 mm. long, filiform. Spikelets 10–18 mm. long, usually brown. Glumes unequal; the inferior (2)3–4·5 (6) mm. long, ovate-oblong; the superior 5–9 mm. long, lanceolate-oblong to narrowly oblong. Florets usually 4–5 (sometimes 3, 6 or 10). Inferior lemma 9–13 mm. long, linear-lanceolate to linear-oblong, acute to acuminate, with the margins involute; fertile lemmas 9–12 mm. long, narrowly lanceolate, acuminate; callus 1–1·5 mm. long, with the hairs 6–10 mm. long. Paleas 1·5–4 mm. long. Anthers c. 4 mm. long. Caryopsis c. 1·5 mm. long.

Botswana. N: Ngamiland, Sowa Pan, 20.iii.1962, *de Beer & Yalala* 14 (BM; SRGH). **Zambia.** N: Chibutubutu, Lukulu R. (Kasama), 1200 m., 16.iv.1958, *Vesey-FitzGerald* 1662 (SRGH). S: Mumbwa, Shibuyunje, 1.viii.1963, *van Rensburg* 2391 (BM; SRGH). **Rhodesia.** N: Sebungwe, Binga Distr., Kowira Hot Springs Camp, 2000, vi.1958, *Davies* 2477 (K). E: Mt. Nuza, 1710 m., 25.vi.1934, *Gilliland* 499b (BM). **Mozambique.** Z: Chinde, between Luabo and Chinde, 13.x.1941, *Torre* 3644 (LISC). SS: Inhambane, Inharrime, Ponta Závora, 4.iv.1959, *Barbosa & Lemos* 8490 (COI; K; LISC; LMA; PRE). LM: Sábiè, margin of R. Incomati, 125 m., v.1893, *Quintas* 126 (COI).

In temperate regions of both hemispheres in the Old World and the New. Recorded from all temperate and subtropical regions of Africa and South Africa. In dense colonies along rivers, in swamps, flood plains, flats, vleis and at the edges of pools or dams, often partly submerged.

2. **Phragmites mauritianus** Kunth, Rev. Gram. **1**: 277, tab. 50 (1830).—C. E. Hubb. in F.T.A. **10**: 155 (1937).—Sturgeon in Rhod. Agric. Journ. **51**: 214 (1954). —C. E. Hubb. in Mem. N.Y. Bot. Gard. **9**, 1: 102 (1954).—Robyns & Tournay in Fl. Parc Nat. Alb. **3**: 202 (1955).—Chippindall in Meredith, Grasses & Pastures of S. Afr.: 229 (1955).—Jackson & Wiehe, Annot. Check List Nyasal. Grass.: 54 (1958).—Conert, Syst. Anat. Arund.: 60 (1961).—W. D. Clayton in F.T.E.A., Gramineae: 120 (1970). TAB. **28**. Type from Mauritius.

Phragmites laxiflorus Steud., Syn. Pl. Glum. **1**: 196 (1854). Type from Ethiopia.

Phragmites vulgaris var *mauritianus* (Bak.) Dur. & Schinz, Consp. Fl. Afr. **5**: 876 (1895) pro parte. Type from Mauritius.

Phragmites vulgaris var. *mossambicensis* (Anderss.) Dur. & Schinz, loc. cit. Type: Mozambique, banks of R. Zambeze, *Peters* (B, †).

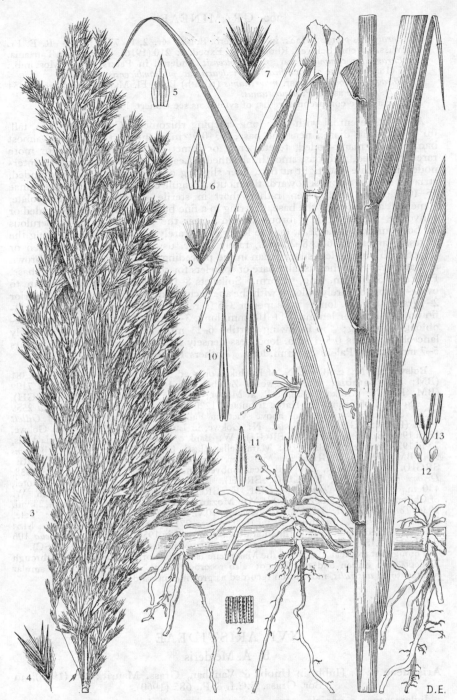

Tab. 28. PHRAGMITES MAURITIANUS. 1, habit (×½) *Milne-Redhead & Taylor* 10540; 2, part of lower surface of leaf-lamina to show scabrousness (×5); 3, inflorescence (×½); 4, spikelet in lateral view (×1½); 5, inferior glume (×4½); 6, superior glume (×4½); 7, florets with glumes removed (×4½); 8, inferior (sterile) lemma (×4½); 9, fertile floret (×1½); 10, fertile lemma (×4½); 11, palea (×4½); 12, lodicules (×4½); 13, sexual organs (×5), all from *Mitchell* 2895.

Phragmites pungens Hack. in Bull. Herb. Boiss., sér. 2, **1**: 771 (1901).—R. E. Fr. in Wiss. Ergebn. Schwed. Rhod.-Kong. Exped. **1**: 210 (1916). Type from Tanzania.
Phragmites communis var. *mossambicensis* Anderss. in Peters, Reise Mossamb.: 555 (1864). Type as for *Phragmites vulgaris* var. *mossambicensis*.
Phragmites communis var. *mauritianus* (Kunth) Bak., Fl. Maurit. and Seych.: 454 (1877). Type as for *P. mauritianus*.
For a more comprehensive list of synonyms see Conert, loc. cit.

A robust perennial with very long creeping rhizomes. Culms (2)4–8 m. tall, up to 4 cm. in diam., near the base, 0·3–0·9 cm. beneath the panicle, almost bamboo-like, many-noded, terete, erect or sometimes floating, simple or more rarely branched, glabrous, smooth, sheathed. Leaf-sheaths longer than the internodes, imbricate and tight at first, later slipping off the culm, dorsally rounded, striate, glabrous or pilose towards the mouth. Ligule c. 1 mm. long. Leaf-laminae 15–100 × 0·75–4 cm. (often rather short in sterile shoots!), linear-lanceolate, somewhat auriculate at the base, tapering to a fine but pungent point, expanded or convolute, rather rigid, glabrous or pilose near the base, distinctly scaberulous on both surfaces or at least on the lower, very rarely smooth, scabrous along the edges. Panicle 15–60 × 8–20 cm., rarely larger, usually rather dense, green or tinged with purple, less silvery than in the preceding species, yellowish or brown at maturity. Branches usually bare of spikelets for some distance from their base. Pedicels 0·5–10 mm. long, filiform. Spikelets 8–16 mm. long. Glumes ovate to ovate-oblong, or rarely oblong, with the apex acute, equal or subequal; the inferior 2–4 (rarely 6) mm. long; the superior 3–5 (rarely 6) mm. long. Florets 4–8(–11)-flowered. Inferior lemma 6–8(–10) mm. long, 3- (rarely 5–7) nerved, lanceolate-oblong to oblong; the following (fertile) 6–11 mm. long, 3–1-nerved, narrowly lanceolate; callus 0·5–1 mm. long, less densely hairy than in *P. australis*; hairs 5–7 mm. long. Paleas 2–6 mm. long. Anthers 1·5–2·25 mm. long.

Botswana. N: c. 40 km. N. of Kachikau on road to Kazane, 10.vii.1937, *Erens* 381 (BM; K). **Zambia.** B: Nangweshi, Zambesi R., 1040 m., 26.vii.1952, *Codd* 7199 (BM; K). N: Mpika, Luangwa R. at Mfuwe, 7.v.1965, *Mitchell* 2895 (BM; SRGH). W: Ndola, Lufanyama R., Kafue tributary, 1220 m., 2.v.1959, *Vesey-FitzGerald* 2559 (BM; SRGH). S: 1·6 km. above Victoria Falls, 6.vii.1930, *Hutchinson & Gillett* 3422a (BM; SRGH). **Rhodesia.** N: Gokwe, in Lutopo vlei c. 32 km. from Gokwe, 18.iii.1963, *Bingham* 535 (BM; SRGH). W: Bank of Zambesi R. above Victoria Falls, 7.viii.1929, *Rendle* 398 (BM). C: Marandellas, Castle Combe, 1430 m., vi.1959, *Wickens* 554 (BM; PRE; SRGH). E: Melsetter, 19.vi.1953, *Crook* 474 (K; LISC; PRE; SRGH). S: Nuanetsi, Bubye R. near Bubye Ranch homestead, c. 550 m., 8.v.1958, *Drummond* 5697 (BM; K; LISC; SRGH). **Malawi.** C: Dowa, Lake Nyasa Hotel, 430 m., 9.viii.1951, *Chase* 3934 (BM; K; SRGH). S: Tangadzi R. 17·6 km. W. of Chiromo, vii.1958, *Seagrief* 3070 (BM). **Mozambique.** N: Sanga, near Sanga, 28.viii.1934, *Torre* 231 (COI; LISC). Z: Mocuba, 10.vi.1943, *Torre* 5458 (LISC). T: Tete, Boroma, Msusa, R. Zambeze, 21.vii.1950, *Chase* 2711 (BM; LISC; SRGH). MS: Chemba, Chieu, Estação Experimental do C.I.C.A., 14.iv.1950, *Lemos & Macuácua* 106 (COI; K; LISC; SRGH). LM: near Lourenço Marques, *Pimenta* 52570 (LISC).
Throughout tropical Africa and the Mascarene Is. with a northern limit passing through Ethiopia, the Sudan and the Congo; also recorded from Natal. Growing in similar habitats to *P. australis*, more often recorded as growing on sandy river banks.

XVI. ARISTIDEAE
By A. Melderis

Aristideae C. E. Hubb. in Hubb. & Vaughan, Grass. Maurit.: 20 (1940); in Bor, Grass. B.C.I. & P.: 685 (1960).

Inflorescence a panicle, open or contracted. Spikelets all alike in shape, 1-flowered; rhachilla disarticulating above the glumes, not produced beyond the floret. Glumes 2, persistent, membranous, usually exceeding the floret, rarely the inferior exceeding the superior, dorsally more or less keeled, 1–5-nerved. Lemma with convolute or involute margins, at first chartaceous, becoming indurated, 1–3-nerved, divided into 3 lobes at the apex, each lobe with its nerve, or only the central lobe, excurrent into an awn, or awns fused with their basal portions into a short beak or a straight or twisted canaliculate column, not articulated or articulated between the apex of the lemma and the base of the awns or the base of the column, or between the apex of the column and the base of the awns, or above the middle of the lemma. Callus obtuse, truncate-emarginate, 2-fid or acuminate, barbate or rarely glabrous. Palea 2-nerved, much shorter than the lemma, oblong, hyaline. Lodicules 2 or 0. Stamens 2 or 1. Ovary glabrous; styles 2, distinct, short; stigmas plumose. Caryopsis tightly embraced by the lemma; hilum linear, more than $\frac{1}{2}$ the length of the caryopsis; starch grains compound. Ligule a ciliate rim. Chromosomes small; basic number 11.

A tribe of 3 genera, distributed mainly in subtropical regions, occupying drier areas with relatively high winter temperatures.

Awns of the lemma scabrid or smooth:
    Glumes 3–5-nerved; awns of the lemma spirally contorted at their base   27. **Sartidia**
    Glumes usually 1-nerved, rarely 3–5-nerved; awns of the lemma suberect or more or
       less spreading, sometimes slightly curved at their base -   -   -   28. **Aristida**
Central awn of the lemma plumose, or if glabrous then with a penicil of hairs at the base
    of the column -   -   -   -   -   -   -   -   29. **Stipagrostis**

27. SARTIDIA de Winter

Sartidia de Winter in Kirkia, **3**: 137 (1963); Bothalia, **8**: 381 (1965).

Spikelets 1-flowered, pedicelled, in terminal laxly contracted panicles. Rhachilla disarticulating above the glumes. Floret ☿, exceeding the glumes. Glumes persistent, narrow, acute or acuminate, 3–5-nerved, awned or awnless. Lemma narrow, subcylindric, somewhat dorsally compressed, slightly tapering upwards, with involute margins, 3-nerved, becoming indurated at maturity, without an articulation, bearing at the base an acute, densely but shortly barbate callus, awned from the apex of the lemma or from the apex of a thick twisted column (if present); awns 3, glabrous or scabrid, rigid, spirally contorted at the base when mature. Palea very small, scale-like, indurated at the base, 2-nerved, glabrous. Lodicules 2, slightly longer than the palea. Stamens 3. Ovary glabrous; styles free; stigmas 2, plumose. Caryopsis dorso-ventrally compressed, deeply grooved, tightly enclosed by the lemma, free, with a linear hilum; embryo $\frac{1}{5}$–$\frac{1}{4}$ the length of the caryopsis. Perennial, densely caespitose grasses with erect, usually simple culms and long linear leaves; ligule a rim of long or short cilia. Panicles erect, narrow, more or less densely contracted, usually interrupted.

A genus of 3 spp. occurring in the Congo, Angola, Zambia, Rhodesia, SW. Africa and S. Africa.

In the external morphology of the spikelets this genus resembles the genus *Aristida*, but differs from the latter in having awns spirally contorted at the base, a caryopsis deeply grooved ventrally, an embryo $\frac{1}{5}$–$\frac{1}{4}$ the length of the caryopsis, and a different leaf anatomy (chlorenchyma of irregular cells forming a continuous tissue between the vascular bundles).

Sartidia angolensis (C. E. Hubb.) de Winter in Kirkia, **3**: 137 (1963); in Bothalia, **8**: 384, fig. 153 & 154 (1965).—Launert in Merxm., Prodr. Fl. SW. Afr. **160**: 165 (1970). TAB. **29**. Type from Angola.
    Aristida vanderystii sensu Stent & Rattray in Proc. Rhod. Sci. Assoc. **32**: 48 (1933).—Sturgeon in Rhod. Agric. Journ. **51**: 506 (1954) non De Wild. 1919.
    Aristida angolensis C. E. Hubb. in Kew Bull. **1949**: 359 (1949). Type as for *Sartidia angolensis*.

Perennial up to 1–2 m. high, densely caespitose, with numerous intervaginal innovations. Culms erect, c. 2 mm. in diam., simple, very firm, solid, terete, glabrous, striate. Leaf-sheaths lax, smooth, often somewhat glossy, estriate in the lower part, finely puberulous and striate upwards. Ligule a dense rim of short cilia; auricles barbate. Leaf-laminae up to 35 × 0·2–0·4 cm., pilose up at the base, densely scaberulous above, finely puberulous below. Panicle up to 40 cm. long, rigid, erect, narrow, slightly contracted, with few spikelets; axis nearly glabrous, striate, compressed, branches solitary or 2-nate, smooth or nearly smooth, bearing 1–few spikelets towards the ends. Spikelets pale-brown to pale-green. Glumes subequal, narrowly linear-lanceolate, chartaceous below, with a membranous acute apex; the inferior 2·5–3·3 cm. long, 5-nerved; the superior c. 2·5 cm. long, 3-nerved. Lemma 1·8–2·2 cm. long (excluding callus and awns),

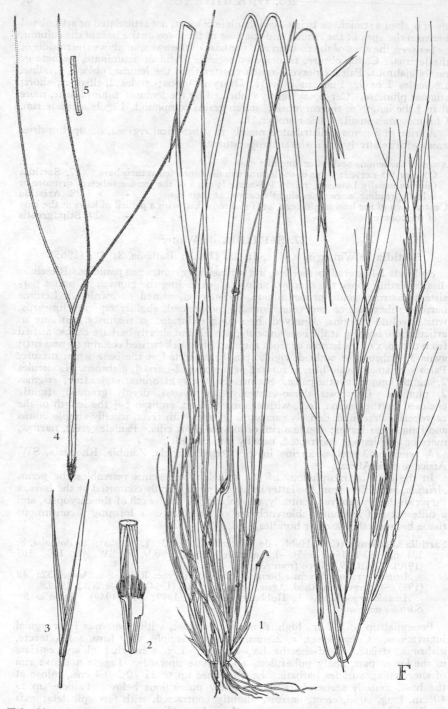

Tab. 29. SARTIDIA ANGOLENSIS. 1, habit (×½); 2, ligule (×6); 3, glumes (×3); 4, lemma (×3); 5, section of awn (×9), all from *Astle* 2493.

linear, coriaceous, glabrous, very finely scaberulous, margins membranous and involute in the lower part; callus 3·5–4mm. long, elongate, acute, densely and shortly barbate; column absent; awns 3, subequal, 7·5–10 cm. long, scaberulous, flattened and slightly grooved on the inner surface. Palea c. 3 mm. long, oblong. Caryopsis 9 mm. long, linear, dorso-ventrally compressed, grooved, dark-brown.

Zambia. B: Mongu Airport, 29.iii.1964, *Verboom* 1352 (K). S: Choma, along Sinazongwe road, 26.v.1963, *Astle* 2493 (K; SRGH). **Rhodesia.** W: Umgusa Spur, Matundhlamahla vlei, 9.iii.1931, *Pardy* 39 (SRGH).
Also in north-western SW. Africa and Angola. In woodlands and burnt places in woodlands, usually on sandy soils.

Differs from *S. vanderystii* (De Wild.) de Winter, which has been found in the Congo Republic, in having a lemma up to 1·8–2·2 cm. long, the column absent, branches of the panicle smooth or nearly smooth, and leaf-sheaths smooth or nearly smooth, often somewhat glossy, striate in the lower part.

28. ARISTIDA L.

Aristida L., Sp. Pl. **1**: 82 (1753); Gen. Pl. ed. 5: 35 (1754).

Spikelets 1-flowered, narrow, pedicelled, in terminal lax or contracted panicles. Rhachilla disarticulating obliquely above the glumes. Floret ♀, equalling or shorter than, or exceeding the glumes. Glumes persistent, equal or unequal, narrow, usually 1–3(5) or 3–5-nerved, acute to acuminate or obtuse, or emarginate, mucronate or awn-tipped. Lemma narrow, terete or laterally compressed, tapering upwards, with convolute or involute margins, 3-nerved, becoming indurated at maturity, bearing at the base an acute, or obtuse, or truncate, or 2-fid, or emarginate, hard, usually barbate callus, awned from the apex of the lemma or from the apex of a usually twisted column (prolongation of the lemma); articulation absent or present (at the apex of the lemma or at the apex of the column); awns 3 (1 central, 2 lateral) or rarely 1, scabrous or glabrous. Palea shorter than the lemma, not keeled, nerveless or 2-nerved, or absent. Lodicules usually 2, rarely 3, obtuse or absent. Stamens 1 or 3. Ovary glabrous; styles free; stigmas 2, plumose. Caryopsis terete or slightly compressed (rarely shallowly grooved), tightly enclosed by the lemma, usually free, with a linear hilum; embryo usually $\frac{1}{3}$–$\frac{1}{2}$ the length of the caryopsis. Annual or perennial, mostly densely tufted, usually with erect or branched, usually solid, rigid culms and long, narrow leaf-laminae, having a ligule of a rim of cilia, and dense spiciform or effuse panicles.

A genus of c. 270 spp., widespread mainly in the warmer tropical and subtropical regions of both hemispheres, occurring also in the temperate regions of N. and S. America.

Henrard in his monograph on *Aristida* (1929–1933) treated this genus in a wider sense, classifying its species into 7 sections (*Schistachne* Henrard, *Stipagrostis* (Nees) Trin. & Rupr., *Pseudochaetaria* Henrard, *Arthratherum* Reichenb., *Pseudarthratherum* Chiov., *Streptachne* Domin and *Chaetaria* Trin.). De Winter in his study on the S. African *Stipeae* and *Aristideae*, based on anatomical, cytological and morphological evidence has split *Aristida* into 3 genera: (1) *Aristida* L. sensu stricto, in which he included Henrard's sections *Chaetaria*, *Pseudochaetaria*, *Arthratherum*, *Pseudarthratherum* and *Streptachne*; (2) *Stipagrostis* Nees, consisting of Henrard's sections *Stipagrostis* and *Schistachne* and (3) *Sartidia* de Winter, separated from the section *Chaetaria* (see pp. 95 and 131).

According to de Winter (1965), these genera can be distinguished by the anatomy of the leaf-laminae, the size and structure of the embryo, by the awns and other organographic characteristics.

Thus, in the leaf-anatomy representatives of *Aristida* sensu stricto differ from those of both other genera in the presence of chloroplasts in both sheaths of the vascular bundles. The outer bundle sheath in *Aristida* sensu stricto consist of smaller (or at least not larger) cells than the inner bundle sheath, and the cells in both bundle sheaths are thin-walled, whilst in *Stipagrostis* and *Sartidia* only the outer bundle sheath contains chloroplasts and the cells of the inner bundle sheath are strongly lignified. The chlorenchyma *in Aristida* sensu stricto is radially arranged around the vascular bundles in a single row of cells, as in

D

Stipagrostis. In *Sartidia* the chlorenchyma cells are continuous between bundles and diffusely, not radially arranged around bundles, forming layers of several cells thick. In *Aristida* sensu stricto silicified cells of the epidermis are usually distinctly dumb-bell like in shape, in *Stipagrostis* usually subcircular in shape, but in *Sartidia* rather variable in size and shape.

The caryopsis in *Aristida* sensu stricto is similar to that of *Stipagrostis*, being terete or slightly compressed and shallowly grooved ventrally, with an embryo ⅓–½ the length of the caryopsis, whilst in *Sartidia* the caryopsis is always dorsoventrally compressed and deeply grooved ventrally, with an embryo up to ¼ the length of the caryopsis.

The awns in *Aristida* sensu stricto are glabrous, scabrid or smooth, but in *Stipagrostis* they are plumose, or the central awn plumose and the lateral ones glabrous, scabrid or smooth, or if the central is glabrous and single, then there is a penicil of hairs at the base of the column. *Sartidia* differs from *Aristida* sensu stricto in having thick, rigid, spirally contorted, glabrous but scabrid awns.

In the present treatment of the Aristidas of our area I have followed de Winter in recognizing *Aristida* sensu stricto, *Stipagrostis* and *Sartidia*.

Most of the species of *Aristida* in our area are characteristic of the drier parts, or of poorer or even skeletal soils, especially when overgrazed. They give very poor grazing to game or cattle. Perhaps because of their affinity for poor soils, they are frequently tolerant of toxic soils and so are characteristic of serpentine and chrome soils (Wild, H. in Kirkia, 5: 49–86 (1965)), copper soils (Wild, H. op. cit. 7: 1–71 (1968)), and nickel soils (Wild, H. op. cit. 7, Suppl.: 1–62 (1970)).

1. Lemma without an articulation - - - - - - - - - 2
— Lemma with an articulation - - - - - - - - - 19
2. Perennials with hard, not easily compressible culm-internodes; lemma beakless or with a beak or twisted column - - - - - - - - - 3
— Annuals or short-lived perennials with soft easily compressible culm-internodes; lemma beakless or without twisted column - - - - - - - 12
3. Lemma beakless - - - - - - - - - - 4
— Lemma with a beak or a twisted column - - - - - - - 6
4. Inferior glume exceeding the superior; spikelets clustered at the ends of branches
1. *hispidula*
— Inferior glume not exceeding the superior; spikelets not clustered at the ends of branches - - - - - - - - - - - - 5
5. Plant with creeping rhizomes; marginal nerves of the leaf-laminae prominent and not much raised above the others; lemma sometimes with 1–2 additional awn-like or setiform appendages; awns of the lemma 30–35 mm. long, strongly curved at the base, divaricately deflexed, delicate - - - - - - 2. *denudata*
— Plant without rhizomes; marginal nerves of the leaf-laminae prominent and much raised above the others; awns of the lemma 7–15 mm. long, slightly spreading, not delicate - - - - - - - - - 3. *canescens*
6. Marginal nerves of the leaf-laminae prominent, much raised above the others; lower leaves flat, usually spirally curved when old, with scattered long hairs above - 7
— Marginal nerves of the leaf-laminae prominent but not much raised above the others; leaves convolute, usually not spirally curved - - - - - - 8
7. Internodes of the culms villous, especially towards the base of the plant; inferior glume slightly exceeding or slightly shorter than the superior; lemma usually with a short twisted column - - - - - - - - - 4. *leucophaea*
— Internodes of the culms glabrous and smooth; inferior glume c. ½ the length of the superior; lemma with a beak - - - - - - - 5. *aemulans*
8. Lemma 10–15 mm. long; panicle linear, spiciform or slightly interrupted at the base
6. *stenostachya*
— Lemma less than 10 mm. long; panicle lax - - - - - - 9
9. Glumes up to 10 mm. long (including the awn), unequal, the inferior c. ½–⅔ the length of the superior - - - - - - - - - - 10
— Glumes usually more than 10 mm. long (including the awn), equal or subequal - 11
10. Lemma 3·5–5 mm. long, with a short beak; awns of the lemma 5–8 mm. long, usually slightly spirally contorted at the base; base of the plant enclosed in fibres from the old leaf-sheaths - - - - - - - - 7. *recta*
— Lemma more than 5 mm. long, with a short beak or a twisted column up to 2·5 mm. long; awns of the lemma 6·5–20 mm. long, suberect or more or less spreading; base of the plant without fibres - - - - - - - 8. *junciformis*
11. Glumes with usually recurved apices; culms branched usually in the upper part; leaf-sheaths glabrous - - - - - - - - 9. *aequiglumis*

— Glumes with apices not recurved; culms branched in the lower part; lower leaf-sheaths hirsute in the upper part, with spreading or flexuous hairs from minute tubercles - - - - - - - - - 10. *textilis*

12. Spikelets up to 3 mm. long (excluding awns) - - - - - 13
— Spikelets more than 5 mm. long (excluding awns) - - - - 14
13. Lemma with 1 awn, 7–10 mm. long - - - - - 11. *diminuta*
— Lemma with 3 awns, of which the central is 4·5–6 mm. long - 12. *cumingiana*
14. Inferior glume not exceeding the superior in length - - - - 15
— Inferior glume exceeding the superior in length - - - - - 17
15. Leaf-sheaths glabrous; auricles without long hairs; branches of the panicle with spikelets from the base; lemma not distinctly tapering towards the apex, usually conspicuously exserted from the glumes when mature - - 13. *adscensionis*
— Leaf-sheaths usually with scattered long hairs; auricles with a few long hairs; branches of the panicle naked in the lower part; lemma distinctly tapering towards the apex, not exserted from the glumes, at least the superior glume longer than the lemma - - - - - - - - - - - - 16
16. Glumes glabrous, or nearly glabrous, smooth, the inferior 2-fid, with a very short awn (up to 0·8 mm. long) from the sinus; lemma glabrous, smooth - - 14. *brainii*
— Glumes scabrid or hirtellous, the inferior gradually tapering into an awn, 1·5–3 mm. long; lemma usually with coarse scabridules in rows towards the apex
15. *scabrivalvis*
17. Panicle effuse, with branches naked for most of their length; inferior glume up to 10 mm. long; awns of the lemma slender; internodes of the culm glabrous - 18
— Panicle effuse or contracted and much interrupted; spikelets diffusely scattered on the branches; inferior glume usually more than 10 mm. long; awns of the lemma rigid; internodes of the culms retrorsely scabrid - - - - 18. *rhiniochloa*
18. Short-lived perennials; lemma c. 6 mm. long, not exserted from the glumes
16. *bipartita*
— Annuals; lemma 10·5–11 mm. long, conspicuously exserted from the glumes
17. *wildii*
19. Column of the lemma absent; articulation between the apex of the lemma and the base of the awns; callus of the lemma obtuse - - - - 19. *hordeacea*
— Column of the lemma present; articulation between the apex of the lemma and the base of the column or between the apex of the column and the base of the awns 20
20. Articulation between the apex of the lemma and the base of the column; callus of the lemma 2-fid or acute - - - - - - - - - - 21
— Articulation between the apex of the column and the base of the awns; callus acute to obtuse - - - - - - - - - - - - 26
21. Callus of the lemma 2-fid - - - - - - - - - - 22
— Callus of the lemma acute, not 2-fid - - - - - - - - 25
22. Auricles of the leaves pubescent or glabrous, not floccose 20. *diffusa* subsp. *burkei*
— Auricles of the leaves floccose (with a tuft of woolly hairs) - - - - 23
23. Lower internodes of the culms pubescent to woolly; column of the lemma 2–7 mm. long - - - - - - - - - - - 21. *vestita*
— Lower internodes of the culms glabrous; column of the lemma usually more than 7 mm. long - - - - - - - - - - - - 24
24. Perennials; column of the lemma usually not exceeding 2 cm. in length
22. *meridionalis*
Annuals; column of the lemma usually exceeding 2 cm. in length 23. *stipoides*
25. Culms with the lower internodes woolly or densely tomentose - 24. *mollissima*
— Culms with the lower internodes glabrous or minutely scaberulous, never woolly or tomentose - - - - - - - - - - - 25. *stipitata*
26. Plant robust, up to 150 cm. tall; panicle contracted but much branched, usually with slightly spreading branches, never spiciform; column of the lemma 1·5 mm. long - - - - - - - - - - - - 26. *pilgeri*
— Plant slender, up to 75 cm. tall; panicle dense, spiciform or effuse, with clusters of the spikelets at the ends of the branches; column of the lemma more than 1·5 mm. long - - - - - - - - - - - - - 27
27. Panicle dense, spiciform or somewhat interrupted, with short branches, usually bearing spikelets to the base - - - - - - - 27. *congesta*
— Panicle effuse, with spreading branches, the spikelets in spiciform clusters at the ends of the branches - - - - - - - - 28. *barbicollis*

1. **Aristida hispidula** Henrard, Monogr. Aristida 2 in Meded. Rijks-Herb. **58a**: 195, t. 89 (1932); Crit. Rev. Aristida Suppl. in Meded. Rijks-Herb. **54c**: 720 (1933).— Stent & Rattray in Proc. Rhod. Sci. Ass. **32**: 46 (1933).—Sturgeon in Rhod. Agric. Journ. **51**: 505 (1954).—Pilg. in Engl. & Prantl, Nat. Pflanzenfam. ed. 2, **14d**: 121 (1956). Type: Rhodesia, Matopos Distr., 1350 m., 14.ii.1930, *Rattray* 16 (L, holotype; K, SRGH, isotypes).

Perennial, up to 70 cm. high, caespitose. Culms erect, simple, glabrous or slightly scaberulous, 2–3-noded; nodes glabrous, smooth. The lowermost leaf-sheaths compressed and keeled, strongly nerved, the upper ones subterete, slightly keeled, scaberulous, tight, sometimes slightly gaping at the apex. Ligule a short ciliate rim; auricles shortly pubescent and with a few long hairs. Leaf-laminae up to 8 cm. long, rather stiff, flat with inrolled margins, conduplicate or convolute, obtusely pointed, with prominent white midrib and marginal nerves, scaberulous or puberulous above, glabrous and smooth beneath. Panicle 10–15 cm. long, long-exserted, with a glabrous, smooth axis; branches 1–6 cm. long, very distant, solitary or 2-nate, at first erect, afterwards more or less spreading, naked in the lower ½, scaberulous or minutely pubescent, with shortly pubescent glandular patches in the axils. Spikelets reddish-green to yellowish-green, densely congested, imbricate, forming small, elongate, spiciform panicles at the ends of the branches. Glumes unequal, linear or nearly lanceolate; the inferior 9–10 mm. long, 1- or nearly 3-nerved (with short lateral nerves), minutely pubescent on the back and scabrid on the keel, gradually tapering into a mucro; the superior 6–8·5 mm. long, 1-nerved, smooth on the keel, slightly scaberulous towards the apex. Lemma 5–5·5 mm. long, fusiform, slightly keeled, glabrous, smooth at the base, with rows of scabridules towards the apex, slightly tapering upwards but scarsely beaked; callus c. 5 mm. long, very obtuse, densely long-barbate; column absent; awns slightly unequal, erect, scabrous; the central awn 8–11 mm. long, the lateral ones slightly shorter; articulation absent.

Rhodesia. N: Darwendale near Rhochrome Mine, Great Dyke, *Wild* 6489 (BM; SRGH). W: Nymandhlovu, Capt. Barry's farm, 24.ii.1932, *Rattray* 498 (BM; SRGH). S: Lundi Reserve, Tweefontein Great Dyke, 17.iii.1964, *Wild* 6413 (BM; SRGH).
Not known from elsewhere. In chrome-rich grassland, chrome vlei and black-land paddocks.

Closely related to *A. bipartita* from which it can be distinguished by the type of inflorescence, having remote, more or less spreading branches, naked in the lower ½, with densely congested, imbricate spikelets, forming elongate secondary spiciform panicles at their ends. In addition the lemmas of *A. hispidula* are glabrous, smooth at the base and with rows of scabridules from c. the middle towards the apex. The panicle of *A. bipartita* is c. ½ the length of the culm, with longer flexuous more delicate branches bearing spikelets more diffusely arranged at the ends of the branches. The lemma of *A. bipartita* is nearly glabrous, smooth, or minutely scaberulous towards the apex.

2. **Aristida denudata** Pilg. in R. E. Fr., Wiss. Ergebn. Schwed. Rhod.-Kongo Exped. 1911–1912, 1: 206 (1916).—Henrard, Crit. Rev. Aristida 1 in Meded. Rijks-Herb. **54**: 135 (1926); Monogr. Aristida 2 in Meded. Rijks-Herb. **58a**: 176, t. 75 (1932).
Syntypes: Zambia, Bangweulu, near Kamindas, in fairly moist meadow, v. 1911, *Fries* 887; on banks of the Kalungwisi R. (between Bangweulu and the Lake Tanganyika), 30.x.1911, *Fries* 1161 (UPS, Syntypes).

Perennial, c. 50 cm. high, caespitose, with well-developed rhizomes, bearing fascicles of culms. Culms erect, slender, simple, glabrous, smooth, with a very long uppermost internode, 1–2-noded; nodes glabrous, smooth. Leaf-sheaths glabrous, smooth, usually shorter than the internodes. Ligule an obsolete minutely ciliolate rim; auricles of the innovations usually long-barbate, those of the culm-leaves dark-coloured, usually glabrescent; collar glabrous or minutely pubescent. Leaf-laminae 10 × 0·1–0·2 cm., flat, acuminate, scabrous or hirtellous above, glabrous, smooth beneath, those of the innovations short, convolute, setaceous, acute. Panicle 5–12 cm. long (excluding awns), lax, with glabrous, smooth rhachis and branches; branches 2-nate, terete, the lower being 4 cm. long, naked in the lower ½, few-flowered, the upper bearing 1–2 spikelets on long filiform, smooth pedicels. Glumes unequal, 1-nerved, glabrous, smooth, acuminate, smooth on the keels, the inferior 8–8·5 mm. long; the superior 10–11 mm. long. Lemma 6·5–7·5 mm. long, glabrous, smooth in the lower part, with rows of minute scabridules towards the apex, having sometimes 1–2 additional awn-like or setiform appendages; callus 0·5 mm. long, obtuse or nearly truncate, shortly barbate; beak or column absent; awns subequal, scaberulous, at first erect, soon divaricately deflexed, the central one 3–3·5 cm. long, more strongly deflexed than the lateral ones which are 2·5–3·0 cm. long; articulation absent.

Zambia. B: Mongu, c. 38 km. NE. of Mongu, 10.xi.1959, *Drummond & Cookson* 6325 (SRGH). N: Mpika Distr., Lake Chibakabaka, 15.x.1963, *Robinson* 5757 (K; SRGH). C: Broken Hill, 23.ix.1947, *Brenan* 7948 (K). **Rhodesia.** W: Matobo, Besna Kobila Farm, 1460 m., i.1960, *Miller* 7106 (SRGH). C: Marandellas, Digglefold, 2.xi.1948, *Corby* 187 (SRGH). **Malawi.** S: Chambe Plateau, Mlanje, 16.xi.1949, *Wiehe* 327 (NYAS; SRGH).
Not known from elsewhere. In shallow depressions, and seasonally inundated swamps.

This is a distinct sp. characterized by the presence of rhizomes bearing fascicles of culms, a lax panicle of few-flowered branches, unequal, glabrous glumes and by a lemma having 1–2 additional awn-like or setiform appendages, as well as divaricately deflexed awns, which are 2·5–3·5 cm. long.

3. **Aristida canescens** Henrard, Monogr. Aristida 2 in Meded. Rijks-Herb. **58a**: 210, 309, t. 95 (1932); Crit. Rev. Aristida Suppl. in Meded. Rijks-Herb. **54c**: 708 (1933).—Stent & Rattray in Proc. Rhod. Sci. Ass. **32**: 46 (1933).—Schweickerdt in Bothalia, **4**: 139 (1941).—Sturgeon in Rhod. Agric. Journ. **51**: 506 (1954).— Chippindall in Meredith, Grasses & Pastures of S. Afr.: 308, map 41 (1955).— de Winter in Bothalia, **8**: 260, fig. 31, 32, 33, 159/10 (1965). Type from S. Africa (Transvaal).

Perennial, up to 1 m. high, densely caespitose. Culms erect, simple or branched, terete or slightly compressed, glabrous or slightly scaberulous below the nodes, c. 4-noded; nodes glabrous. Leaf-sheaths tight, glabrous or scaberulous, the lower slightly compressed. Ligule a ciliolate rim; auricles barbate or slightly pubescent; collar minutely pubescent. Leaf-laminae usually up to 20 cm., scarcely 1 mm. wide when expanded, coarse, involute or flat at the base, curling when old, scabrid or hirtellous above, glabrous or scaberulous beneath, with thickened scabrous margins. Panicle up to 20 cm. long, erect, lax or contracted but much interrupted, with an angular or triquetrous scabrous axis; branches remote, naked in the lower part, scabrid. Spikelets pallid or yellowish-green, sometimes tinged with purple, densely congested towards the ends of the branchlets. Glumes unequal, lanceolate, 1-nerved; the inferior 5·5–8 mm. long, scaberulous on the keel, subobtuse or emarginate or slightly 2-fid, with very obtuse lateral lobes, mucronate from the sinus; the superior 8–11 mm. long, smooth, truncate at the apex and shortly awned or 2-lobed, with a mucro from the sinus. Lemma 7–11 mm. long, slightly compressed, smooth or scaberulous on the keel; callus 1 mm. long, subobtuse, densely barbate; beak or column absent; awns scabrous, subequal, the central 9–15 mm. long, the lateral ones 7–13 mm. long; articulation absent. $2n = 55$.

Botswana. SE: Towani, 15.iv.1931, *Pole Evans* 3215 (SRGH). **Zambia.** S: Between Kafue and Mazabuka on Munali Pass, 15.iii.1963, *van Rensburg* 1571 (BM; SRGH). **Rhodesia.** N: Lomagundi, Caledonian Ranch, 1370 m., 12.iv.1959, *Phipps* 2162 (BM; SRGH). W: Nymandhlovu, 24.ii.1932, *Rattray* 497 (BM; SRGH). C: Selukwe, 24 km. S. of Selukwe on Great Dyke, 16.iii.1964, *Wild* 6366 (BM; SRGH). E: Umtali, 1020 m., iv.1920, *Perrott* in Eyles no. 3068 (K). S: Gwande Ratanhyana, 11.vi.1968, *Wild* 7730 (BM; SRGH).
Also in the Cape Prov., Natal, Orange Free State and the Transvaal. On sandy, gravelly soils, in open grassland, woodland and sandveld.

Resembling *A. junciformis* but differing from the latter in having terete culms, and leaf-laminae with prominent and raised marginal nerves, coarser spikelets, and in the absence of beak or column.

4. **Aristida leucophaea** Henrard, Crit. Rev. Aristida 2 in Meded. Rijks-Herb. **54a**: 298 (1927); Monogr. Aristida 2 in Meded. Rijks-Herb. **58a**: 163, t. 65 (1932).— Stent & Rattray in Proc. Rhod. Sci. Ass. **32**: 46 (1933).—Sturgeon in Rhod. Agric. Journ. **51**: 504 (1954).—W. D. Clayton, F.T.E.A., Gramineae: 143 (1970). Type: Rhodesia, Salisbury, hillside plot, 1440 m., v.1920, *Eyles* 2238 (holotype, K; isotype, SRGH).
Aristida eriophora Henrard, Monogr. Aristida 2 in Meded. Rijks-Herb. **58a**: 165, t. 66 (1932); Crit. Rev. Aristida Suppl. in Meded. Rijks-Herb. **54c**: 716 (1933).—Sturgeon in Rhod. Agric. Journ. **51**: 504 (1954). Type: Rhodesia, Hatfield, 1470 m., 11.iv.1931, *Stent* in GHS. 3976 (L, holotype; K, SRGH, isotypes).

Perennial, 30–80 cm. high, caespitose. Culms erect, simple, or slightly branched from some of the upper nodes, densely greyish- or whitish-woolly, sometimes in the upper part less woolly or pubescent only, 5–many-noded; nodes usually glabrous. Leaf-sheaths tight or lax, somewhat keeled, glabrous or minutely pubescent or more or less pilose. Ligule a ciliolate rim; auricles pubescent or barbate; collar glabrous or pubescent. Leaf-laminae up to 20 cm. long, with prominent or raised marginal nerves, scabrous and more or less pilose above, scaberulous or glabrous, smooth beneath, laminae of the lower leaves c. 3 mm. wide, flat, curled when old, those of the upper scarcely 1 mm. wide when expanded, convolute, setaceously acuminate. Panicle 5–15 cm. long, long-exserted, contracted and more or less interrupted, with scaberulous or minutely pubescent or more or less pilose axis and branches; branches single or 2-nate, angular, appressed, much branched, the lowermost 5–6 cm. long, naked at the base, with a shorter few-flowered branchlet at the base. Spikelets yellowish-green or brownish, sometimes tinged with purple. Glumes equal or slightly unequal, linear-lanceolate, usually scabridulous, sometimes slightly pilose, keeled, 1-nerved; the inferior 3·5–8 mm. long, scabrous on the keel, gradually tapering into an awn up to 3 mm. long, the superior 5–8 mm. long, sometimes nearly glabrous on the back, usually glabrous on the keel, 2-fid at the apex, with an awn c. 1 mm. long from the sinus. Lemma 5–6 mm. long, glabrous, smooth at the base, densely scaberulous towards the apex; callus c. 0·5 mm. long, obtuse, densely short-barbate; column up to 2·5 mm. long, slightly twisted, scabrous; awns unequal, suberect or slightly spreading; the central 1·1–1·6 mm. long, the lateral ones 9–11 mm. long; articulation absent.

Zambia. B: Senanga Distr., Marshi R. fringe, 30.ix.1964, *Verboom* 1134 (K; SRGH). C: Chilanga, Agriculture Research Station, 1260 m., 17.iv.1953, *Hinds* 87 (K). S: Mazabuka, Mapangazia paddock, 1200 m., 3.v.1963, *Astle* 2365 (K; SRGH). **Rhodesia.** N: Trelawney, Tobacco Station, 17.iv.1943, *Moffett* 114 (SRGH). W: Shangani, Gwampa Forest Reserve, iii.1955, *Goldsmith* 152 (SRGH). C: Marandellas, Grasslands, 7.vii.1947, *Newton* 88 (SRGH). E: Inyanga, Sawunyama Reserve, 1070 m., 17.vi.1958, *Plowes* 2041 (BM; COI; LISC; SRGH). **Mozambique.** SS: Muchopes Manjacaze, 25.vi.1944, *Torre* (LISC).
Not known elsewhere. On sandy soils or schists in grassland, on grassy hills, in open woodland and on edges of vleis.

A distinct sp. which can readily be recognized by densely woolly culms, strongly curled old leaves having prominent and much-raised marginal nerves, a contracted and more or less interrupted panicle, slightly unequal, usually densely scaberulous glumes and by lemmas with a short beak or column up to 2·5 mm. long. In the presence of the woolly culms it resembles *A. vestita* and *A. mollissima*, but the latter are more robust plants which differ from *A. leucophaea* in having auricles with a tuft of woolly hair, a different type of the inflorescence, unequal glumes and in a longer callus of the lemma which is 2-fid (in *A. vestita*) or acuminate (in *A. mollissima*). In addition the lemmas of *A. vestita* and *A. mollissima* have an articulation between the apex of the lemma and the base of the column. As indicated by Henrard, *A. leucophaea* should differ from *A. eriophora* in having inverse glumes (the inferior exceeding the superior in its length) and by the presence of a short column, whilst in *A. eriophora* the inferior glume is shorter than the superior and the lemma possesses a beak. When characteristics of the spikelets of isotypes of these spp. were compared critically, it was found that the characters mentioned above are very variable on the same plant and cannot be used for diagnostic purposes. There are also no essential differences in the vegetative characters of both spp., such as the pubescence of culms and the structure and pubescence of the leaves. The two spp., therefore, are treated here as being conspecific.

5. **Aristida aemulans** Melderis in Bol. Soc. Brot., Sér. 2, **44**: 285, t. III, IV, fig. C (1970). TAB. **30.** Type from Congo.

Perennial, 40–70 cm. high, caespitose. Culms erect, branched at the base, glabrous, smooth, 2–3-noded; nodes glabrous, smooth. Leaf-sheaths lax, somewhat keeled, glabrous, smooth. Ligule a densely short-ciliate rim; auricles barbate; collar glabrous or slightly pubescent. Leaf-laminae of the lower leaves up to 40 × 0·2–0·25 cm., flat, spirally curled, when old densely nerved, with prominent, flattened, much-raised marginal nerves, scabrid on the nerves, usually with scattered long hairs above, glabrous and smooth beneath. Panicle 13–20 cm. long, long-exserted, contracted, more or less interrupted in the lower part, with a scabrid axis; branches usually single, angular, much-branched, appressed, scabrous, the lowermost 3–5 cm. long, naked at the base. Spikelets purplish-green or brownish,

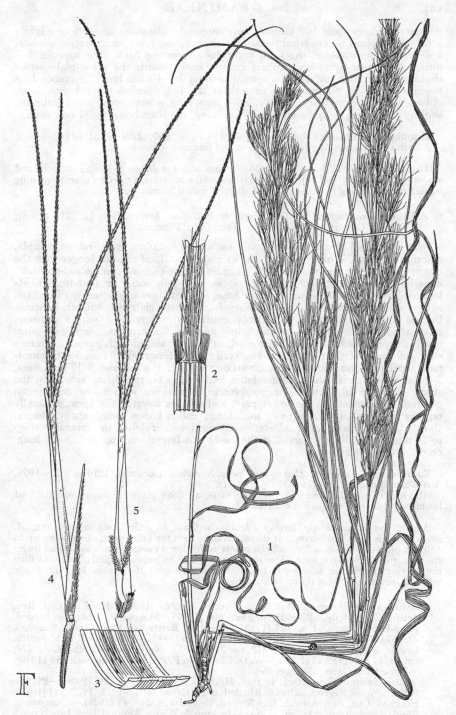

Tab. 30. ARISTIDA AEMULANS. 1, habit ($\times\frac{1}{2}$); 2, ligule ($\times 6$); 3, section of leaf-lamina ($\times 18$); 4, spikelet ($\times 18$); 5, lemma ($\times 18$), all from *Gathy* 1165.

with scaberulous pedicels. Glumes very unequal, subobtuse or acute or slightly 2-fid, with a short mucro from the sinus; the inferior 3·5–4 mm. long, lanceolate, 1–3-nerved, scabrous on the keel, scabridulous on the back; the superior 6–7 mm. long, linear-lanceolate, scabrid on the keel towards the apex, glabrous or slightly scaberulous towards the apex. Lemma 3·5–4 mm. long, ± scaberulous towards the apex, glabrous and smooth at the base; callus 0·3–0·4 mm. long, obtuse, densely short-hairy; beak c. 0·5–1 mm. long, scaberulous; awns scabrous, slightly spreading; central awn 1·2–1·5 cm. long, the lateral ones 0·8–1 cm. long.

Zambia. N: Chisenga Ranch, Luwingu Distr., ix.1960, *Astle* 30 (SRGH). Also in the Congo. At edges of streams and dambos.

This sp. is closely allied to *A. leucophaea* from which it differs in having glabrous and smooth culms, an inferior glume c. ⅔ the length of the superior, a superior glume equalling the lemma (including its beak) and in a shortly beaked lemma.

6. **Aristida stenostachya** W. D. Clayton in Bol. Soc. Argent. Bot. **12**: 111 (1968); F.T.E.A., Gramineae: 143 (1970). Type from Tanzania.

Perennial, 60–130 cm. high, densely caespitose. Culms erect, robust, simple, glabrous, smooth, 4–many-noded; nodes glabrous. Leaf-sheaths longer than the internodes, gaping, glabrous, smooth. Ligule a short-ciliolate rim; auricles usually densely long-barbate; collar glabrous, shortly pubescent or barbate. Leaf-laminae 20–60 × 0·2–0·5 cm., flat at the base, involute upwards, densely ribbed (all ribs of equal thickness), slightly scaberulous above, glabrous, smooth beneath, Panicle 25–40 cm. long, dense, narrow, contracted, interrupted, with a glabrous terete striate axis; branches erect, slightly angular, scaberulous, bearing spikelets nearly from the base. Spikelets greenish or slightly tinged with purple. Glumes subequal, punctulate on the sides, 1-nerved; the inferior 10–19 mm. long, acuminate, tapering into a short awn, scabrous on the keel; the superior 9–17 mm. long, 2-fid at the tip, with short obtuse lobes and a very short delicate awn from the sinus. Lemma 10–15 mm. long, subterete, sometimes with a few dark spots, glabrous and smooth in the lower part, scaberulous towards the apex, gradually tapering into a short beak up to 4 mm. long; callus 1 mm. long, narrow, obtuse, shortly barbate; awns subequal, somewhat rigid, scaberulous; the central awn up to 30 mm. long, slightly curved at the base; the lateral ones up to 25 mm. long, erect; articulation absent.

Zambia. N: Abercorn Distr., lane near Abercorn Common, 1710 m., 2.iv.1959, *McCallum Webster* A259 (K; LISC; SRGH).
Also in Kenya and Tanzania. On sandy soils, on flood plains, in open woodland and bushland, clearings, and in old cultivated land.

Among the perennial spp. having a lemma without an articulation and column, *A. stenostachya* is most distinctive. It is more robust than the other spp., often being up to 130 cm. high and differing from all the others in having a narrow dense somewhat interrupted panicle, 25–40 cm. long. It is also characterized by large subequal glumes of which the inferior exceeds the superior in length, and by a lemma 10–15 mm. long, gradually tapering into a short beak up to 4 mm. long.

7. **Aristida recta** Franch. in Bull. Soc. d'Autun, **8**: 365 (1896).—Henrard, Crit. Rev. Aristida **2** in Meded. Rijks-Herb. **54a**: 500 (1927); Monogr. Aristida **2** in Meded. Rijks-Herb. **58a**: 261, t. 125 (1932).—Stent & Rattray in Proc. Rhod. Sci. Ass. **32**: 47 (1933).—Schweickerdt in Bothalia, **4**: 148 (1941).—Chippindall in Meredith, Grasses & Pastures of S. Afr.: 312, fig. 277 (1955).—de Winter in Bothalia, **8**: 270, fig. 46, 47 & 48, 159/15 (1965).—W. D. Clayton in F.T.E.A., Gramineae: 145 (1970). —Type from Gabon (Franceville).
Aristida atroviolacea Hack. in Bull. Herb. Boiss , Sér. 2, **6**: 707 (1906).—Pilg. in R.E.Fr., Wiss. Ergebn. Schwed. Rhod.-Kongo Exped. 1911–1912, **1**: 205 (1916).— Henrard, Crit. Rev. Aristida **1** in Meded. Rijks-Herb. **54**: 45 (1926).—Sturgeon in Rhod. Agric. Journ. **51**: 507 (1954).—Jackson & Wiehe, Annot. Check List Nyasal. Grass.: 29 (1958). Type from S. Africa (Natal).
Aristida gossweileri Pilg. in Engl., Bot. Jahrb. **39**: 598 (1907). Type from Angola.
Aristida hockii De Wild. in Bull. Jard. Bot. Brux. **6**: 39, pl. 35 (1919).—Henrard, Crit. Rev. Aristida **2** in Meded. Rijks-Herb. **54a**: 237 (1927). Type from the Congo (Katanga).

Perennial 10–30 cm. high, densely caespitose. Culms erect, very elegant, slightly compressed, simple, glabrous, smooth, 1-(2) noded; nodes glabrous, smooth. Leaf-sheaths striate, glabrous, the basal somewhat lax, persistent, breaking up into fibres, forming a dense tuft at the base of the culms, the upper somewhat tighter. Ligule a short-ciliolate rim; auricles barbate; collar glabrous. Leaf-laminae up to 20 × 0·1 cm., setaceously involute, scaberulous above, glabrous, smooth beneath, subobtuse or acute. Panicle 8 × 4 cm., ovate-lanceolate in outline, long-exserted, lax but not much branched; axis smooth, more or less angular, scaberulous upwards; branches 2-nate or 3–5-partite, capillary, somewhat flexuous, naked in the lower ½, scaberulous, the lower up to 3 cm. long. Spikelets brownish-purple to deep purple, contracted towards the ends of the branches. Glumes sub-equal to unequal, glabrous or scabrous on the keels; the inferior usually 2·5–5 mm. long, lanceolate-ovate, 3-nerved, often 3-keeled, gradually or abruptly tapering into a short awn c. 1 mm. long; the superior usually 4·5–6 mm. long, linear-lanceolate, mucronate, or with a short awn up to 1 mm. long. Lemma 3·5–5 mm. long, slightly keeled, glabrous, smooth in the lower part, more or less scaberulous towards the apex, prolonged into a short beak up to 0·7 mm. long; callus 0·2–0·3 mm. long, obtuse, short-barbate; awns subequal, scabrid, usually slightly contorted at the base; the central 7–8 mm. long, the lateral ones 5–6 mm. long; articulation absent.

Zambia. B: Mankoya, 56 km. on Mongu road, 8.xi.1959, *Drummond & Cookson* 6243 (BM; COI; SRGH). N: Kasama Distr., Mungwi, 8.x.1960, *Robinson* 3918 (K; SRGH). W: Kitwe, 12.x.1957, *Fanshawe* 3778 (SRGH). C: Rufunsa to Lusaka, 7.ix.1947, *Greenway & Brenan* 8073 (BM; K). E: Lundazi, xi.1961, *Verboom* 504 (BM). S: Choma, 14.ix.1957, *Robinson* 2437 (PRE; SRGH). **Rhodesia.** N: Sipolilo, 14.v.1959, *Chapman* 506 (SRGH). C: Salisbury, 1460 m., ix.1919, *Eyles* 1795 (K; SRGH). E: Melsetter, Lionhills Forest Reserve, 1550 m., 5.x.1956, *Crook* 574 (K; SRGH). **Malawi.** N: 3·2 km. W. of Chisenga, 1950 m., 27.viii.1962, *Tyrer* 596 (BM; SRGH). C: Dedza Mt., 6.ix.1950, *Jackson* 168 (K; SRGH). S: Mlanje, Chame Plateau, 2130 m., 16.xi.1949, *Wiehe* n/330 (K; SRGH). **Mozambique.** Z: Gúruè, Summit Serra da Gúruè, 25.ix.1944, *Mendonça* 2283 (LISC).

Also from the Guinea region to the Congo Republic, Uganda, Tanzania, Transvaal and Swaziland. Usually on damp ground in dambos, sometimes on exposed ridges, also in seepage areas in acid mountain grassland.

A distinct perennial sp. characterized by having 10–30 cm. high slender culms with dense fibrous tufts of old leaf-sheaths at the base, a long-exserted lax panicle (ovate-lanceolate in outline) with small purple or brown spikelets. Its lemma prolonged into a short beak bearing short awns which are usually slightly spirally contorted at the base.

8. **Aristida junciformis** Trin. & Rupr., Sp. Gram. Stip.: 143 (1842).—Steud., Syn. Pl. Glum. **1**: 140 (1854).—Dur. & Schinz, Consp. Fl. Afr. **5**: 804 (1894).—Stapf in Harv. & Sond., F.C. **7**: 556 (1899).—Henrard, Crit. Rev. Aristida **2** in Meded. Rijks-Herb. **54a**: 273 (1927); Monogr. Aristida **2** in Meded. Rijks-Herb. **58a**: 287, t. 140 (1932).—Schweickerdt in Bothalia, **4**: 141 (1941) pro parte.—Chippindall in Meredith, Grasses & Pastures of S. Afr.: 309, map 42 (1955), pro parte.—Pilg. in Engl. & Prantl, Nat. Pflanzenfam. ed. 2, **14d**: 121 (1956).—de Winter in Bothalia, **8**: 266, fig. 41, 42, 43, 159/14 (1965), pro parte.—W. D. Clayton, F.T.E.A., Gramineae: 144 (1970).—Launert in Merxm., Prodr. Fl. SW. Afr. **160**: 32 (1970). Type from S. Africa.

Perennial, up to 60 cm. high, densely caespitose, with old dark leaf-sheaths at the base and with thick spongy roots. Culms erect, slender, wiry, simple or slightly branched, glabrous, smooth, 3–4-noded; nodes glabrous, smooth. Leaf-sheaths tight, the lower keeled, striate, glabrous, smooth or scabrous, sometimes with long flexuous hairs. Ligule a short-ciliolate rim; auricles pubescent or more or less barbate, especially those of the lower sheaths. Leaf-laminae up to 30 × 0·1 cm. when expanded, subsetaceous, convolute, mostly strictly erect, sometimes slightly curved, scaberulous or hirtellous above, glabrous smooth beneath. Panicle 5–20 × 1–8 cm., narrow, contracted or fairly lax, erect or somewhat nodding, with a more or less angular, scaberulous axis; branches up to 5 cm. long, usually 2-nate, erect, scaberulous, subfascicled, capillary, often naked in the lower part, with spikelets congested towards the ends. Glumes linear-lanceolate, unequal, thin, acute or acuminate or awned, 1-nerved (the inferior sometimes with 2 additional nerves in the lower part), glabrous or minutely pubescent. Lemma c. 6 mm.

long, glabrous, smooth or scaberulous towards the apex; callus c. 0·5 mm. long, shortly barbate; awns very fine, suberect, more or less spreading; articulation absent. 2n = 44.

A very variable sp. consisting of several subspp., of which subsp. *macilenta* and subsp. *welwitschii* occur in the F.Z. area. The subsp. *junciformis*, according to Henrard, is distributed in Natal and the Orange Free State. De Winter has recorded *A. junciformis* also from the Cape Prov., SW. Africa and the Transvaal. As he has treated this sp. in a wider sense, also referring to it forms having a well-developed column of the lemma, the distribution of *A. junciformis*, given by him, is in need of a revision. The subsp. *junciformis*, which has not yet been recorded from the F.Z. area, is characterized by having internodes of the culms of more or less equal length, being strongly compressed especially below, by very rigid, convolute, mostly strictly erect leaf-laminae, by inferior glume 6–7 mm. long, by a superior glume 9–10 mm. long and by a beakless or nearly beakless lemma with somewhat unequal awns (the central 20–35 mm. long, the lateral ones 15–28 mm. long). Both the other subspp. have terete or only slightly compressed internodes of the culms below the nodes bearing branches. Their leaf-laminae are sometimes flat or folded at the base and less rigid than in subsp. *junciformis*.

Culms 20–70 cm. high; internodes of the culms unequal; lemma with a short beak up to 1 mm. long; awns very unequal (central 9–15 mm. long, the lateral ones 6·5–10 mm. long) - - - - - - - - - - - subsp. *macilenta*
Culms 70–140 cm. high; internodes of the culms more or less equal; lemma with a twisted column 2–2·5 mm. long; awns subequal, usually 15–20 mm. long
                                                                subsp. *welwitschii*

Subsp. **macilenta** (Henrard) Melderis in Bol. Soc. Brot., Sér. 2, **44**: 287 (1970). Type: Malawi, 1895, *Buchanan* 238 (BM, holotype); paratype: Rhodesia, Salisbury, 1440 m., iii.1920, *Eyles* 2142 (K, paratype; SRGH, isoparatype).
    Aristida macilenta Henrard, Crit. Rev. Aristida **2** in Meded. Rijks-Herb. **54a**: 319 (1927); Monogr. Aristida **2** in Meded. Rijks-Herb. **58a**: 178, t. 76 (1932).—Sturgeon in Rhod. Agric. Journ. **51**: 504 (1954).—Jackson & Wiehe, Annot. Check List Nyasal. Grass.: 29 (1958). Type as above.
    Aristida contractinodis Stent & Rattray in Proc. Rhod. Sci. Ass. **32**: 47 (1933). Type: Rhodesia, Salisbury, S. of Makabusi R., 1470 m., iii.1920, *Eyles* 1592 (SRGH, holotype).

Culms 20–70 cm. high. Internodes of the culms very unequal; the lower short, the following very long (10 cm. or more), the upper very short (sometimes less than 1 cm.), the uppermost long-exserted. Leaf-laminae usually spreading at more or less right angles from the culm, those of the uppermost leaves often approximate in pairs. Auricles of the leaves especially of the lower ones, long-barbate. Inferior glume usually 2·5–5 mm. long, superior one 4·5–7 mm. long. Lemma c. 5·5 mm. with a short beak, usually up to 1 mm. long; awns unequal (the central usually 9–11 mm. long, the lateral ones 6·5–8 mm. long).

Zambia. C: Serenje, Kundalila Falls, viii.1968, *Williamson* 1114 (BM). S: 19·2 km. N. of Choma, Muckle Neuk, 1280 m., 27.xi.1954, *Robinson* 976 (SRGH). **Rhodesia.** N: Gokwe, 0·9 km. N. of Gokwe, 11.iii.1963, *Bingham* 597 (BM; SRGH). W: Matobo, Farm Besna Kobila, 1460 m., xi.1956, *Miller* 3926 (BM; LISC; SRGH). C: Marandellas, Digglefold, 17.x.1948, *Corby* 157 (SRGH). E: Inyanga, Gairezi Ranch on Mozambique border, 9·6 km. N. of Troutbeck, 1370 m., 13.xi.1956, *Robinson* 1871 (SRGH). S: Fort Victoria, Makoholi Experimental Farm, 8.i.1948, *Robinson* 189 (SRGH). **Malawi.** C: Lilongwe, Ciledzi, Cankhnadwe dambo, 24.ix.1950, *Jackson* 170 (SRGH).
    Also in the Transvaal, Tanzania and Kenya. On sandy and serpentine soils, often in damp situations, in grassland, dry and burnt dambo, wet sandveld, vlei, edges of dry vleis and near streams between rocks.

    It is closely allied to *A. transvaalensis* from which it differs mainly in having simple or slightly branched culms and in a beaked column of the lemma. In *A. transvaalensis* culms are much branched and the lemma possesses a twisted column, usually c. 2 mm. long, sometimes up to 6 mm. long. Some specimens from our area referred to by Henrard to *A. transvaalensis*, e.g. Rhodesia, Rusape, ii.1931, *Fitt* 39 (SRGH), are in my opinion depauperate forms of subsp. *welwitschii*.

Subsp. **welwitschii** (Rendle) Melderis in Bol. Soc. Brot., Sér. 2, **44**: 287 (1970). Type from Angola.

Aristida welwitschii Rendle, Cat. Afr. Pl. Welw. **2**: 202 (1899).—Henrard, Crit. Rev. Aristida **3** in Meded. Rijks-Herb. **54b**: 682 (1928); Monogr. Aristida **2** in Meded. Rijks-Herb. **58a**: 228, t. 107 (1932). Type as above.

Aristida welwitschii var. *minor* Rendle, tom. cit.: 203 (1899).—Henrard, in tom. cit.: 683 (1928). Type from Angola.

Aristida angustata Stapf in Harv. & Sond., F.C. **7**: 556 (1899).—Henrard, Crit. Rev. Aristida **1** in Meded. Rijks-Herb. **54**: 26 (1926). Type from S. Africa (Cape).

Aristida welwitschii var. *subtomentosa* Henrard, op. cit. 546: 684 (1928). Type from Angola.

Aristida pardyi Stent & Rattray in Proc. Rhod. Sci. Ass. **32**: 47 (1933).—Sturgeon in Rhod. Agric. Journ. **51**: 507 (1954). Type: Rhodesia, Nyamandhlovu, Umgusa Spur, Matundhlamahla Vlei, c. 1050 m., 10.iii.1031, *Pardy* 46 (holotype, SRGH; isotype, K).

Culms 70–140 cm. high. Internodes of the culms more or less equal (the lower short, the following aproximately of equal length). Leaf-laminae usually suberect, those of the uppermost leaves never approximate in pairs. Auricles of the leaves usually pubescent. Inferior glume 5·5–6 mm. long; the superior 6·5- 7 mm. long. Lemma c. 6 mm. long, with a twisted column 2–2·5 mm. long; awns subequal, usually 15–20 mm. long.

Zambia. B: Mongu, 11.iv.1966, *Robinson* 6927 (K). N: Abercorn, Lake Chila, 1680 m., 29.iv.1955, *Siame* 642 (SRGH). N: Ndola, 20.iv.1961, *Fanshawe* 4369 (K). C: Serenje, Lusiwashi dambo near Kanona, c. 1430 m., 6.iv.1961, *Vesey-FitzGerald* 2938 (SRGH). **Rhodesia.** N: Lomagundi, Muriel Mine, Great Dyke, 21.iii.1963, *Wild* 6087 (BM; SRGH). W: Shangani, Gwampa Forest Reserve, iii.1955, *Goldsmith* 153 (SRGH; K). C: Marandellas, Chiota Reserve, 16.i.1959, *Cleghorn* 421 (SRGH). S: Fort Victoria, Glyntor, 14.xi.1947, *Robinson* 38 (SRGH).

Also in Angola, and S. Africa (Cape Prov.). On sandy, black clayey and serpentine soils, usually in wet situations, in grassland, on ungrazed margins of vleis, in sandveld, *Acacia-Mopane* veld and on edges of dambos and termite mounds.

9. **Aristida aequiglumis** Hack. in Bull. Herb. Boiss. **3**: 381 (1895).—Stapf in Harv. &. Sond., F.C. **7**: 555 (1899).—Henrard, Crit. Rev. Aristida **1** in Meded. Rijks-Herb. **54**: 18 (1926); Monogr. Aristida **2** in Meded. Rijks-Herb. **58a**: 236, t. 112 (1932).—Stent & Rattray in Proc. Rhod. Sci. Ass. **32**: 46 (1933).—Schweickerdt in Bothalia, **4**, 1: 145 (1941).—Sturgeon in Rhod. Agric. Journ. **51**: 506 (1954).—Chippindall in Meredith, Grasses & Pastures of S. Afr.: 310 (1955).—de Winter in Bothalia, **8**: 262, fig. 34, 35, 36, 159/11 (1965). Type from S. Africa (Transvaal).

Perennial, usually c. 40 cm. high, densely caespitose. Culms erect, slender, usually simple, rarely branched from the upper nodes, terete, glabrous, smooth or minutely scaberulous below the nodes, c. 3–4-noded; nodes glabrous, smooth. Leaf-sheaths usually tight, glabrous, smooth, persistent. Ligule a short-ciliolate rim; auricles minutely pubescent, sometimes with long hairs; collar minutely pubescent. Leaf-laminae up to $15 \times 0·1$–0·15 cm. (when expanded), erect or slightly curved, usually setaceously convolute, rather firm, minutely puberulous above, glabrous, smooth beneath. Panicle 6–15 cm. long, linear-oblong, usually lax and open, sometimes contracted; axis terete in the lower part, more or less angular upwards, slightly scaberulous; branches up to 4 cm. long, solitary or 2-nate, suberect, filiform, minutely scaberulous, with 1–3 spikelets. Glumes subequal, linear-lanceolate, 1-nerved, with a prominent glabrous or scabrid keel, scabrid or shortly hirtellous on the back, often with somewhat recurved apices; the inferior 5–9 mm. long, with a short awn up to 1·5 mm. long, the superior 5·5–9·5 mm. long, usually 2-fid, with a mucro from the sinus. Lemma 6–7 mm. long, minutely scaberulous towards the apex; callus c. 0·5 mm. long, subobtuse or obtuse, shortly barbate; column 2–8 mm. long, twisted, scabrid; awns somewhat unequal, filiform, suberect or slightly spreading, scaberulous, the central 15–25 mm. long, the lateral ones 13–20 mm. long; articulation absent. 2n = 22.

Zambia. C: Chakwenga Headwaters, 100–129 km. E. of Lusaka, 27.iii.1965, *Robinson* 6554 (K). S: Mapanza, Mt. Makulu Pasture Research Station, 1250 m., 14.iv.1956, *Robinson* 1469 (SRGH). **Rhodesia.** C: Salisbury, 1460 m., 1.iv.1927, *Eyles* 4777 (K; SRGH).

Also in the Transvaal. On sandy ground in open situations, seasonally flooded areas, paddocks, rock crevices and on hill slopes. Considered to be an indicator of eroded soil.

Closely allied to *A. textilis* from which it differs in having shorter leaf-laminae, the superior glume being usually slightly 2-fid with a short mucro from the sinus, a longer lemma and by the absence of long scattered hairs from minute tubercles on the lower leaf-sheaths and on the upper surface of the leaf-laminae.

10. **Aristida textilis** Mez in Fedde, Repert. **17**: 149 (1921).—Henrard, Crit. Rev. Aristida 3 in Meded. Rijks-Herb. **54b**: 631 (1928); Monogr. Aristida 2 in Meded. Rijks-Herb. **58a**: 168, t. 69 (1932).—Jackson & Wiehe, Annot. Check List Nyasal. Grass.: 29 (1958). Type from Tanzania (Kyimbila).

Perennial, c. 50 cm. high, densely caespitose. Culms erect, slender, simple or branched from some of the lower nodes, glabrous, smooth or sometimes slightly minutely pubescent below the nodes, 4–6-noded; nodes glabrous, smooth. Leaf-sheaths usually longer than the internodes of the culm, striate, usually tight, the lower more or less densely hirsute, with spreading or flexuous hairs from minute tubercles, the upper glabrous, smooth. Ligule a very short, ciliolate rim; auricles shortly pubescent or becoming glabrous, sometimes with long hairs; collar minutely pubescent. Leaf-laminae up to 25 × 0·1–0·15 cm. (when expanded), convolute, ending in a fine point but not pungent, sometimes more or less flat towards the base, with prominent marginal nerves, scaberulous or hirtellous (with scattered long hairs from minute tubercles) above, scaberulous or smooth beneath. Panicle mostly 10–15 cm. long, sometimes up to 20 cm., somewhat contracted but not dense, usually spiciform with a minutely scaberulous axis. Branches 3–7 cm. long, 2-nate or simple, erect or suberect, naked in the lower part, or with spikelets from the base, few-flowered. Spikelets usually purple, sometimes greenish tinged with purple. Glumes subequal, keeled, glabrous or minutely scaberulous, more or less gradually tapering into a short awn up to 2 mm. long; the inferior 4–7 mm. long, 1-nerved or indistinctly 3-nerved, the superior 5–8 mm. long, 1-nerved, usually smooth. Lemma c. 5 mm. long, minutely scaberulous towards the apex; callus 0·3–0·5 mm. long, obtuse, densely short-barbate; column 1–2 mm. long, twisted, scaberulous; awns subequal, 13–15 mm. long, delicate, slightly minutely scaberulous, somewhat spreading; articulation absent.

Zambia. N: Abercorn, Sunzu Mt., 32 km. S. of Abercorn, c. 1680 m., 20.iv.1961, *Vesey-FitzGerald* 3309 (BM; SRGH). S: Kafue Gorge, 18.iii.1960, *Vesey-FitzGerald* 2737 (SRGH). **Malawi.** N: Ekwendeni, Zombure (? Zombwe), 19.i.1951, *Jackson* 360 (K; NYAS).
Also in Tanzania. On red sandy loam, mainly in *Brachystegia* woodland, also in shady crevices of rocky hills and in grassland.

This sp., which is closely related to *A. aequiglumis*, is characterized by the much-branched culms in their lower part, by the presence of numerous long leaves, having long hairs from minute tubercles on the lower leaf-sheaths and on the upper surface of the lower leaves and by a slightly exserted, somewhat contracted panicle with purple or purplish spikelets.

11. **Aristida diminuta** (Mez) C. E. Hubb. in Kew Bull. **4**: 480 (1949).—Sturgeon in Rhod. Agric. Journ. **51**: 504 (1954).—Jackson & Wiehe, Annot. Check List Nyasal. Grass.: 29 (1958).—W. D. Clayton in F.T.E.A., Gramineae: 146 (1970). Type: Malawi, 1861, *Buchanan* 561 (B, holotype, K, isotype).
   Stipa diminuta Mez in Fedde, Repert. **17**: 208 (1921).—Type as above.
   Aristida cumingiana var. *reducta* Pilg. in Engl., Notizbl. Bot. Gart. Berl. **11**: 805 (1933). Type from Tanzania.
   Aristida cumingiana var. *uniseta* Stent & Rattray in Proc. Rhod. 'Sci. Ass. **32**: 48 (1933). Syntypes: Rhodesia, Salisbury, iv.1909, *Allen & Nobbs* 746; Salisbury, Borrowdale, 1500 m., v.1927, *Eyles* 4931; Trelawney, v.1931, *Fitt* 165 (syntypes, SRGH; isotypes, K).
   Aristida cumingiana var. *diminuta* (Mez) Jacques-Félix in Journ. Agric. Trop. **13**: 51 (1966). Type as for *Aristida diminuta*.

Annual, 10–30 cm. high, caespitose. Culms erect, slender, simple or branched at the base, glabrous, smooth 1–2-noded; nodes glabrous, smooth. Leaf-sheaths glabrous, smooth, or minutely pubescent, striate, keeled. Ligule a short-ciliate rim; auricles pubescent, with some long hairs; collar glabrous or minutely pubescent. Leaf-laminae 2–7 (up to 10) cm. long, up to 1 mm. wide, very narrow, involute, scabrous and with long scattered hairs above, glabrous, smooth beneath. Panicle usually 4–10 cm. long, usually more than ½ as long as the whole plant, lax, effuse, sometimes contracted; axis scabrid; branches capillary, scaberulous, naked below, the lower 2–3-nate. Spikelets 2·5–3 mm. long, dark purple or greenish tinged with purple. Glumes unequal, glabrous, smooth, or scaberulous towards the apex, keeled, scabrous on the keel, mucronate or shortly awned; the inferior 2–2·5 mm. long, lanceolate, nearly 3-nerved; the superior 2·5–3 mm. long, narrowly lanceolate, 1-nerved. Lemma 1·5–1·7 mm. long, purplish or greenish, scaberulous towards the apex; callus 0·2–0·3 mm. long, broadly obtuse, shortly barbate; beak or column absent; awn 7–10 mm. long, delicate, scabrous, slightly recurved (the lateral awns absent); articulation absent.

Zambia. C: E. of Kashitu, 1370 m., v.1920, Eyles 2849 (K; SRGH). S: Choma, 9.v.1963, Astle 2426 (BM; K; SRGH). Rhodesia. N: Goromonzi, slopes of Ngomo Kurira, 1520 m., 3.iv.1960, Phipps 2792 (SRGH). C: Hartley, Poole Farm, 1220 m., 29.iii.1948, Hornby 2939 (K; SRGH). S: Bikita, W. bank of Turgwe R. at confluence with Dafana R., 1100 m., 5.v.1969, Biegel 3028 (BM; SRGH). Malawi. S: Mlanje Distr., 16. km. NW. of Likabula Forestry Depot, 700 m., 15.vi.1962, Robinson 5354 (K; LISC; SRGH). Mozambique. Z: Montes do Ile, 19.vi.1943, Torre 5518 (LISC). T: Angónia, between Vila Coutinho and the frontier, 15.v. 1948, Mendonça 4150 (LISC).
Also in Tanzania, reported also from Mali and Chad. On damp waste ground, in flushes on granite slopes, sandveld, along drainage courses or margins of vleis or dambos.

A small delicate annual, usually much branched at the base, with an effuse panicle, and with usually dark-purplish spikelets. In general appearance closely resembling A. cumingiana from which it can be distinguished by lemmas having only 1 awn. As the distributional area of A. diminuta is restricted to tropical E. Africa whilst A. cumingiana has a much wider distribution, occurring also in Senegal and Ethiopia and extending eastwards through southern parts of Asia to Philippines, they are treated as separate spp.

12. **Aristida cumingiana** Trin. & Rupr., Sp. Gram. Stip.: 141 (1842).—Henrard, Crit. Rev. Aristida 1 in Meded. Rijks-Herb. 54: 120 (1926); Monogr. Aristida 2 in Meded. Rijks-Herb. 58a: 159, t. 62 (1932).—Stent & Rattray in Proc. Rhod. Sci. Ass. 32: 48 (1933).—Sturgeon in Rhod. Agric. Journ. 51: 504 (1954).—Pilg. in Engl. & Prantl, Nat. Pflanzenfam. ed. 2, 14d: 121 (1956).—Bor, Grass. of B.C.I. & P.: 409 (1960).—W. D. Clayton in F.T.E.A., Gramineae: 146 (1970). FRONTISP. Type from Philippine Is.
Aristida delicatula A. Rich., Tent. Fl. Abyss. 2: 393 (1851). Type from Ethiopia.

Annual, mostly 5–10(25) cm. high, simple or densely caespitose. Culms erect, usually much branched below, glabrous, smooth, 1–2-noded; nodes glabrous, smooth. Leaf-sheaths smooth or minutely pubescent, striate, keeled. Ligule a short-ciliate rim; auricles pubescent or with a few long hairs; collar glabrous or minutely pubescent. Leaf-laminae up to 6 cm. long, scarcely 1 mm. wide, very narrow, involute, scabrous, with scattered long hairs above, glabrous, smooth beneath. Panicles ½ as long as the whole plant or sometimes even longer, lax, or effuse, somewhat contracted; axis glabrous or scabrid; branches capillary, scaberulous, naked below, the lower 2-nate or in 3. Spikelets dark-purple or greenish tinged with purple. Glumes unequal, gradually tapering into a short mucro, glabrous, smooth or scaberulous towards the apex, keeled; the inferior 1·5–2 mm. long, lanceolate, scabrous on the keel, nearly 3-nerved, the superior 2–2·5 mm. long, narrowly lanceolate, smooth on the keel, 1-nerved. Lemma 1·5–2·5 mm. long, greenish or purplish, lanceolate or ovate-lanceolate, glabrous and smooth in the lower part, with rows of minute scabridules towards the apex; callus 0·1–0·2 mm. long, obtuse, shortly barbate; beak and column absent; awns unequal, delicate, scabrid, the central one 4·5–6 mm. long, slightly recurved, the lateral ones 2·5–4 mm. long, suberect; articulation absent.

Zambia. B: Mongu, Lealui, 11.iv.1966, Robinson 6924 (BM; SRGH). N: Kasama-Mporokoso road near Lukupa R., 18.iv.1958, Vesey-FitzGerald 1695 (SRGH) W: Mufulira, 4.v.1934, Eyles 8416 (K; SRGH). C: E. of Kashitu, 1370 m., v.1920, Eyles 2849a (K; SRGH). E: Fort Jameson, 22.vi.1963, Verboom 1110 (BM; K). Rhodesia. N: Sipolilo Distr., Nyarasuswe, 14.v.1962, Wild 5471 (BM; K; SRGH). C: Salisbury, iv.1909, Eyles 6007 (SRGH). Malawi. N: Karonga Distr., Chisenga, 23.v.1962, Robinson 5224 (K; LISC; SRGH).

Also in Tanzania and extending from Senegal and Ethiopia eastwards through India, Siam and China to the Philippines. On yellow-sandy or vlei (serpentine or norite) soils, mainly in open situations, on hillsides, in clearings, semi-dambos in woodland areas and on roadsides in moist dambos.

This delicate annual sp. with small dark purplish spikelets is closely allied to *A. diminuta* from which it differs mainly in having lemmas with 3 awns.

13. **Aristida adscensionis** L., Sp. Pl. **1**: 82 (1753).—Dur. & Schinz, Consp. Fl. Afr. **5**: 799 (1894).—Rendle in Cat. Afr. Pl. Welw. **2**: 202 (1899).—Stapf in Harv. &. Sond., F.C. **7**: 554 (1899).—Henrard, Crit. Rev. Aristida **1** in Meded. Rijks-Herb. **54**: 8 (1926); Monogr. Aristida **2** in Meded. Rijks-Herb. **58a**: 322, t. 157 (1932).— Hutch. in Hutch. & Dalz., F.W.T.A. **2**: 532 (1936).—Sturgeon in Rhod. Agric. Journ. **51**: 508 (1954).—Pilg. in Engl. & Prantl, Nat. Pflanzenfam. ed. 2, **14d**: 120 (1956).—Jackson & Wiehe, Annot. Check List Nyasal. Grass.: 29 (1958).—Bor, Grass. of B.C.I. & P.: 407 (1960); Fl. Iran. **70**: 336 (1970).—W. D. Clayton, F.T.E.A., Gramineae: 148 (1970).—Launert in Merxm., Prodr. Fl. SW. Afr. **160**: 30 (1970). Type from Ascension I.

    Aristida submucronata Schumach., Beskr. Guin. Pl.: 67 (1827).—Henrard, Crit. Rev. Aristida **3** in Meded. Rijks-Herb. **54b**: 609 (1928); Monogr. Aristida **2** in Meded. Rijks-Herb. **58a**: 321, t. 157 (1932).—Schweickerdt in Bothalia, **4**: 152 (1941).—Sturgeon in Rhod. Agric. Journ. **51**: 507 (1954).—Chippindall in Meredith, Grass. & Pastures of S. Afr.: 313 (1955).—Pilg. in Engl. & Prantl, loc. cit. Type from Guinea.

    Chaetaria curvata Nees, Fl. Afr. Austr.: 186 (1841).—Type from S. Africa.

    Aristida vulgaris Trin. & Rupr., Sp. Gram. Stip.: 131 (1842) nom. illegit.

    Aristida guineensis Trin. & Rupr., op. cit.: 137 (1842).—Steud., Syn. Pl. Glum. **1**: 139 (1854). Type from Guinea.

    Aristida thonningii Trin. & Rupr., loc. cit.—Henrard, Crit. Rev. Aristida **3** in Meded. Rijks-Herb. **54b**: 632 (1928). Type from Guinea.

    Aristida mauritiana A. Rich., Tent. Fl. Abyss. **2**: 392 (1851) nom. illegit. non Kunth (1829). Type from Ethiopia.

    Aristida curvata (Nees) Dur. & Schinz, Consp. Fl. Afr. **5**: 802 (1894).—Henrard, Crit. Rev. Aristida **1** in Meded. Rijks-Herb. **54**: 124 (1926); Monogr. Aristida **2** in op. cit. **58a**: 318, t. 156 (1932).—Schweickerdt, tom. cit.: 149 (1941).—Chippindall in Meredith, Grasses & Pastures of S. Afr.: 312, fig. 279, map 43 (1955).—de Winter in Bothalia, **8**: 253, fig. 25, 26, 159/7 (1965). Type as above.

    Aristida adscensionis subsp. *guineensis* (Trin. & Rupr.) Henrard, Crit. Rev. Aristida **1** in Meded. Rijks-Herb. **54**: 216 (1926).—Schweickerdt in Bothalia, **4**: 152 (1941). —de Winter in Bothalia, **8**: 250, fig. 21, 159/5 (1965). Type as for *Aristida guineensis*.

    Aristida curvata var. *nana* Henrard, Crit. Rev. Aristida **3** in Meded. Rijks-Herb. **54b**: 487 (1928); Monogr. Aristida **2** in Meded. Rijks-Herb. **58a**: 318 (1932). Type from S. Africa.

    Aristida submucronata var. *scabra* Henrard, Crit. Rev. Aristida **3** in Meded. Rijks-Herb. **54b**: 610 (1928); Monogr. Aristida **2**, op. cit. **58a**: 321 (1932). Type: Mozambique, Boroma, *Menyhart* 916 (Z, holotype).

    Aristida adscensionis var. *guineensis* (Trin. & Rupr.) Henrard, Monogr. Aristida **2** in Meded. Rijks-Herb. **58a**: 325, t. 159 (1932).—Chippindall in Meredith, Grasses & Pastures of S. Afr.: 313 (1955).—Robyns & Tournay, Fl. Parc Nat. Alb. **3**: 136 (1955). Type as for *Aristida guineensis*.

Annual, 10–100 cm. high, caespitose. Culms erect, somewhat geniculately ascending, often branched from the base and lower nodes, glabrous, smooth, often purplish; nodes glabrous, smooth. Leaf-sheaths tightly or laxly embracing the culm, more or less keeled, striate, smooth or scaberulous. Ligule a short-ciliate rim; auricles smooth or minutely pubescent; collar glabrous. Leaf-laminae up to 15 × 0·1–0·25 cm., linear, usually flat, scabrid above, usually glabrous beneath, with margins conspicuously thickened. Panicle up to 25 × 0·1–0·25 cm., exserted, erect, narrow and dense, interrupted at the base, or lax, many-flowered with a scabrous axis and appressed, erect or somewhat spreading, scabrous or scaberulous branches. Spikelets greenish or green tinged with purple. Glumes unequal, linear-lanceolate to lanceolate-oblong, more or less keeled; the inferior 4–7 mm. long, 1-nerved, minutely scaberulous or glabrous, emarginate at the apex with obtuse lobes and with an exserted mucro from the sinus, or acute, scabrous or scaberulous on the keel; the superior up to 6–8·5 mm. long, erose, 2-fid at the apex, with a mucro or short awn from the sinus, becoming scabrous on the keel towards the apex. Lemma (5·5)8–10(14) mm., usually exceeding, sometimes as long as, the glumes, tubular or slightly compressed, not narrowed upwards, pallid, usually with purple spots or tinged with purple, finely punctulate and scabrous on the keel, rarely scabrous on lateral sides towards the apex; column absent; awns unequal or subequal, more or less winged towards the base, very scabrous, the central awn being up to 20 mm. long, lateral ones 8–15 mm. long; callus 0·5–0·7 mm. long, obtuse, densely shortly barbate; articulation absent. 2n =22.

Botswana. N: Lake Ngami, near Sehitwa, 24.iii.1961, *Vesey-FitzGerald* 3323 (BM; K; SRGH). SW: 9·6 km. SW. of Takatshwane on road to Lehututu, 21.ii.1960, *de Winter* 7423 (K; LISC). **Zambia.** N: Kasakalwe, Abercorn, Lake Tanganyika, 775 m., 15.v.1960, *Vesey-FitzGerald* 2792 (SRGH). C: Mt. Makulu Research Station, 14·4 km. S. of Lusaka, 30.v.1957, *Angus* 1613 (SRGH). E: Fort Jameson, i.1962, *Verboom* 543 (BM; SRGH). S: Mazabuka, iv.1932, *Trapnell* 1077 (K; SRGH). **Rhodesia.** N: Bindura, 26.iii.1932, *Simpson Bros.* 5640 (SRGH). W: Matobo, Matopos Research Station, 1960, *Rattray* 1880 (SRGH). C: Salisbury, 1460 m., ii.1920, *Reynolds* 2465 (K; SRGH). E: Chipinga, Lower Sabi, east bank 1·6 km. NE. of Hippo Mine, 370 m., 12.iii.1957, *Phipps* 569 (COI; LISC; SRGH). S: Glyntor, Fort Victoria, 27.xii.1947, *Robinson* 59 (K; SRGH). **Malawi.** C: Linthipe Valley, 28.iii.1950, *Wiehe* 473 (SRGH). S: Port Herald, between Thangadzi and Lilenje R., c. 90 m., 25.iii.1960, *Phipps* 2698 (BM; SRGH). **Mozambique.** N: Malema, Mutuáli, c. 174 m., 12.ii.1964, *Correia* 131 (COI; LISC). T: Posto Zootécnico de Angónia, 12.v.1948, *Mendonça* 4183 (LISC). MS: Chemba, Chiou, Estação Experimental do C.I.C.A., 20.iv.1960, *Lemos & Macuácua* 125 (BM; COI; K; LISC; LMA; SRGH). SS: Limpopo, near to Posto de Chigubo, 12.iii.1951, *Barbosa & Myre* 88 (K; LMA). LM: Maputo, near to Umbelúzi Experimental Station bank of R. Umbelúzi, 8.iv.1949, *Myre* 511 (LISC; LMA).

Widely distributed in the warm and hot regions of the Old and New Worlds. On sandy, serpentine, basalt, paragneiss, dolorite, dark and sandy clay, loam, and red-clayey soils in *Mopane* veld, sandveld, open grassland, savannah, open coastal scrub, scrub-woodland, on stony hills, termite mounds, alluvial ground along river banks, dry lake-bed flats, beach, along roadsides, in cultivated fields and on disturbed ground.

This sp. has a very wide geographical distribution, and exhibits a considerable variation in its external morphology. It has therefore been divided by some authors into several species, subspecies or varieties, which appear to represent the more distinct variants of the numerous populations. Of species recognized by Henrard (1932), the following occur in our area: (1) *A. adscensionis*, particularly its variety var. *guineensis* (characterized by an open panicle with naked branches at the base, glumes glabrous on their lateral sides, an acute inferior glume and terete or subtriquetrous awns of the lemma without broad hyaline margins towards their base); (2) *A. submucronata* (panicle erect and rather contracted, spiciform, somewhat interrupted at the base, having branches crowded with spikelets nearly to the base; glumes minutely scaberulous on their lateral sides, the inferior acute; awns of the lemma robust, triquetrous, with a distinct mid-rib and hyaline rather broadly winged margins) and (3) *A. curvata* (similar to *A. submucronata*, but with its panicle narrower and more interrupted, the inferior glume with a very obtuse, erose 2-fid apex and a short mucro from the sinus and more slender and less-winged awns of the lemma). These differential characters, however, are very variable. Some authorities have, therefore, considered *A. submucronata* to be identical with *A. adscensionis*. De Winter (1965) has, with some reservations, kept *A. adscensionis* var. (" subsp.") *guineensis* apart from *A. curvata*. According to him, *A. adscensionis* var. *guineensis* can be separated from *A. curvata* by the presence of smaller spikelets, a more open panicle and by lemmas usually as long as, or slightly shorter than the glumes or rarely exceeding them. (It should be mentioned that the holotype of *A. adscensionis*, preserved in the Linnaean Herbarium, London, possesses similar glumes.) He has indicated a close relationship between both these taxa and the necessity for re-examination of the whole complex, involving *A. adscensionis*, *A. submucronata* and *A. curvata*. Examinations of the specimens from our area have shown that none of these alleged differences in the characters given by Henrard or de Winter appears to be constant. *A. adscensionis* var. *guineensis*, *A. submucronata* and *A. curvata* intergrade morphologically and have overlapping geographical distributions. It is, therefore, not possible to maintain them as separate species. Further analysis, by modern methods is required before a comprehensible and consistent subdivision of this sp. is possible.

14. **Aristida brainii** Melderis in Bol. Soc. Brot., Sér. 2, **44**: 279, t. I, IV, fig. A (1970). Type: Rhodesia, Victoria Falls, 10–15.iii.1932, *Brain* 8866 (K, holotype; SRGH, isotype).

    Aristida serrulata sensu Stent & Rattray in Proc. Rhod. Sci. Ass. **32**: 46 (1933).— Sturgeon in Rhod. Agric. Journ. **51**: 505 (1954).

Annual, up to 50 cm. high, delicate. Culms erect, ascending, usually branched from the lower nodes, glabrous, smooth, 3–4-noded; nodes glabrous and smooth. Leaf-sheaths rather lax, keeled; the lower ones with some scattered long, delicate hairs; the upper ones glabrous and smooth. Ligule densely short-ciliolate; auricles ciliolate, often with a few long hairs; collar glabrous. Leaf-laminae up to 16 × 0·17 cm., densely ribbed, with marginal nerves and midrib more prominent, shortly pubescent on the nerves above, glabrous and smooth beneath. Panicle up to 20 cm. long, effuse; axis terete and glabrous, smooth in the lower part, subangular, scabrid in the upper; branches 1–2 at the nodes, subangular, slightly scabrid, divaricate, naked in the lower part, bearing 1–3 spikelets, with dark, shortly pubescent glandular patches in the axils. Spikelets delicate, greenish or purplish-brown, the lateral ones divaricate, on short pedicels (up to 1·5 mm. long). Glumes broad, unequal, 2-fid at the apex, 1-nerved, keeled; the inferior 6–7·5 mm. long, scabrid on the keel, otherwise glabrous and smooth, rarely slightly minutely pubescent towards the apex, with obtuse lobes at the apex with a very short awn (up to 0·8 mm. long) from a shallow sinus, exceeding the lobes; the superior 8·5–10 mm. long, nearly smooth on the keel, with acute lobes at the apex and with a very short awn (up to 0.5 mm. long) from a deep sinus, usually not exceeding the lobes. Lemma c. 8 mm. long, keeled, glabrous and smooth, sometimes mottled with purple, deeply grooved ventrally, with enrolled margins; callus c. 1 mm. long, subobtuse, densely pubescent; column absent; awns subequal, 1–1·5 cm. long, slightly dilated towards the base, scabrid; articulation absent.

Rhodesia. W: Victoria Falls, 910 m., iii.1932, *Brain* 5830 (SRGH). Not known from elsewhere. Ecological details not available.

This species has been confused with *A. serrulata*, described by Chiovenda from Eritrea, which is closely allied to *A. rhiniochloa*. *A. brainii* differs from *A. serrulata* in the following diagnostic characters: its culms (including nodes) and leaf-sheaths are glabrous, smooth (with a few long scattered hairs on the lower leaf-sheaths and auricles); the branches of the panicle are glabrous, conspicuously spreading; the spikelets are smaller; glumes glabrous, or nearly glabrous, smooth, with the inferior shorter than the superior; and the awns of the lemma shorter (10–15 mm. long). *A. brainii* is closely related to *A. scabrivalvis*, from which it differs in shorter branches of the panicle, glabrous glumes, and a 2-fid inferior glume with a very short awn (up to 0·8 mm. long) from the sinus; its lemma is glabrous.

15. **Aristida scabrivalvis** Hack. in Bull. Herb. Boiss., sér., 2, **6**: 708 (1906).—Henrard, Crit. Rev. Aristida **3** in Meded. Rijks-Herb. **54b**: 534 (1928); Monogr. Aristida **2** in Meded. Rijks-Herb. **58a**: 202, t. 91 (1932).—Stent & Rattray in Proc. Rhod. Sci. Ass. **32**: 46 (1933) pro parte.—Schweickerdt in Bothalia, **4**: 137 (1941).—Sturgeon in Rhod. Agric. Journ. **51**: 505, 506 (1954).—Chippindall in Meredith, Grasses & Pastures of S. Afr.: 308, map 40 (1955).—Pilg. in Engl. & Prantl, Nat. Pflanzenfam. ed. 2, **14d**: 121 (1956).—De Winter in Bothalia, **8**: 254, fig. 27, 28, 29, 159/8 (1965).—W. D. Clayton in F.T.E.A., Gramineae: 147 (1970).—Launert in Merxm., Prodr. Fl. SW. Afr. **160**: 33 (1970). Type from S. Africa (Transvaal).
     Aristida bipartita sensu Stent & Rattray in Proc. Rhod. Sci. Ass. **32**: 46 (1933).

Annual, up to 85 cm. high, caespitose. Culms erect or ascending, slender, usually glabrous, smooth, much branched, 3–4-noded; nodes glabrous, smooth. Leaf-sheaths laxly embracing the culm, keeled, glabrous or minutely scaberulous, the lower often with some scattered long hairs. Ligule a short-ciliolate rim; auricles densely and shortly ciliolate, often with a few long hairs, or nearly glabrous; collar glabrous. Leaf-laminae 30 × 0·1–0·35 cm., usually flat, fairly rigid with 1–2 pallid prominent marginal nerves, scaberulous and hirtellous above, glabrous, smooth, or minutely scaberulous beneath. Panicle 15–30 cm. long, terminal and lateral, usually lax; axis more or less terete, glabrous in the lower part, angular, scabrous in the upper part; branches often filiform and sometimes flexuous, naked in the lower part, compressed, angular, scabrous, with a dark glandular patch in the axils. Spikelets linear, purplish. Glumes subequal to unequal, 1-nerved, scaberulous or hirtellous, with a scabrid keel; the inferior 4–6·5 mm. long, lanceolate, more or less gradually tapering into an awn up to 3 mm. long or with a slightly 2-fid apex, with an awn up to 2·5 mm. long from the sinus; the superior up to 5·5–6·5 mm. long, deeply 2-fid, with an awn up to 1·5 mm. long from the sinus. Lemma 5–7 mm. long, slightly keeled, purple or green, darkly mottled, glabrous or punctulate in the lower part, with rows of coarse scabridules towards the apex, rarely glabrous all over; callus c. 1·5 mm. long, rounded to subacute, densely short-barbate; beak or column absent; awns 10–14 mm. long, subequal, erect or somewhat spreading, very scabrous; articulation absent. 2n = 22.

A distinct species which is characterized by having a large panicle, conspicuously awned, scabrid or hirtellous, subequal or unequal glumes (the superior longer than the inferior), and by a scabrid, rarely nearly glabrous lemma which is subequal or shorter than the glumes.

Panicle open, ovate in outline, with more or less spreading branches:
   Branchlets and pedicels of the spikelets spreading; spikelets effuse, more or less equally
      scattered on the branches or branchlets   -   -   -   - subsp. *scabrivalvis*
   Branchlets and pedicels of the spikelets appressed; spikelets congested at the ends of
      the branches, forming spiciform secondary panicles   -   - subsp. *borumensis*
Panicle narrow, spiciform, with very short contracted branches and branchlets
                                                                  subsp. *contracta*

Subsp. scabrivalvis

Botswana. N: Bushman's Pits, 96 km. E. of Maun, 27.iii.1961, *Vesey-FitzGerald 3343* (SRGH). **Zambia.** N: Mpika Distr., 45·1 km. NW. of Mfuwe, Kapiri Plots, c. 700 m., 21.iii.1969, *Astle 5635* (BM; SRGH). S: Livingstone, 22.iii.1961, *Fanshawe 6447* (BM; SRGH). **Rhodesia.** N: Lomagundi, 1460 m., 25.ii.1959, *Phipps, Drummond & Jackson 1524* (BM; COI; K; SRGH). W: Matopos Research Station, 29.iii.1961, *Kennon MRSH 3248* (SRGH). C: Sebakwe R. near Que Que, 6.iii.1961, *Vesey-FitzGerald 3103* (BM; SRGH). E: Umtali, Premier Estate, 1130 m., 7.ii.1925, *Eyles 4213* (BM; SRGH). S: Fort Victoria, Glyntor, 19.i.1948, *Robinson 205* (K; SRGH).

With glabrous lemmas. **Rhodesia.** S: Gwanda, Todd's Hotel, 20.iii.1949, *West 2890* (SRGH).

Also in the Cape Prov., SW. Africa, Natal, the Transvaal and Tanzania. On heavy and sandy soils, red basalt loam, serpentine and red dolorite in sandy pockets over loose limestone cobbles, on grassy hills, in open flat grassland, grassy valley bottoms, bush savanna, on banks of rivers, on waste land and disturbed ground.

A. scabrivalvis subsp. *scabrivalvis* resembles *A. bipartita* from which it differs mainly in the annual habit, distinctly awned glumes and by the superior glume exceeding the inferior in length. It is characterized by having a large inflorescence, minutely scabridulous or hirtellous glumes and usually a coarsely scabrid lemma, which is subequal to, or shorter than, the glumes.

Subsp. borumensis (Henrard) Melderis in Bol. Soc. Brot., Sér. 2, **44**: 287 (1970).
   Type: Mozambique, Boroma, xii. 1890, *Menyharth* (L, holotype).
   Aristida borumensis Henrard, Monogr. Aristida **2** in Meded. Rijks-Herb. **58a**: 203, t. 91 (1932); Crit. Rev. Aristida Suppl. in Meded. Rijks-Herb. **54c**: 706 (1933).—Sturgeon in Rhod. Agric. Journ. **51**: 506 (1954). Type as above.
   Aristida filiformis Henrard, Monogr. Aristida **2** in Meded. Rijks-Herb. **58a**: 203, t. 117 (1932); Crit. Rev. Aristida Suppl. in Meded. Rijks-Herb. **54c**: 717 (1933).—Sturgeon in Rhod. Agric. Journ. **51**: 506 (1954). Type: Rhodesia, Wankie, near Fuller, c. 900 m., 11.iii.1929, *Pardy HC. 4934* (SRGH, holotype; K, isotype).
   Aristida scabrivalvis Hack., Stent & Rattray in Proc. Rhod. Sci. Ass. **32**: 46 (1933) pro parte.

Botswana. N: Okavanga, Nokaneng, Taokhe flood plain, 21.iii.1961, *Vesey-FitzGerald 3315* (BM; SRGH). **Zambia.** S: Mapanza N., 1070 m., 9.iv.1955, *Robinson 1222* (K; SRGH). **Rhodesia.** N: Bindura, Leaopard's Vlei, 27.iii.1948, *Rattray 1482* (K; SRGH). W: Wankie Distr., opposite Kazungula, 910 m., iv.1955, *Davies 1058* (K; LISC; SRGH). C: Gatooma, c. 1142 m., 20.iii.1930, *Brain 951* (K; SRGH). E: Melsetter, 1220–1830 m., iv.1907, *Swynnerton 1603* (BM; SRGH). S: Beitbridge, 9·6 km. ENE. of Tuli Police Camp, 23.iii.1959, *Drummond 5966* (BM; K; SRGH). **Malawi.** N: Karonga Distr., iii.1954, *Jackson 1257* (K).

With glabrous lemmas. **Botswana.** SE: 24·8 km. N. of Mahalapye, 15.iv.1931, *Pole Evans 3205* (SRGH). **Zambia.** S: Mazabuka, v.1932, *Trapnell 1070* (SRGH). **Rhodesia.** W: Wankie, c. 900 m., 11.iii.1929, *Pardy 4934* (SRGH).

Not known from elsewhere. On sandy and basalt soils and on clayey loam in "mopane" tree-savanna, in paddocks, along roadsides and on waste or disturbed ground.

Examination of the holotype of *A. filiformis* Henrard (*Pardy 4934*) shows that it represents a slender plant which agrees very well, in nearly all panicle and spikelet characters, with subsp. *borumensis* except for the lemmas which are glabrous. The presence or absence of the scabridity on the lemmas is not of great taxonomic importance since the forms with glabrous lemmas are met with also in the other subsp. As this feature is not accompanied with any other essential diagnostic characters, these forms are not treated here as separate taxa. The absence of long hairs on leaf-sheaths is also not an important character, because the forms with scabrid lemmas, also do not always possess such hairs.

In its very lax panicle with branches bearing a few spikelets at the ends, subsp. *borumensis* resembles *A. bipartita*, from which it could be distinguished mainly by having a superior glume longer than the inferior and by the scabridity or pubescence of the glumes. The lemma of *A. scabrivalvis* subsp. *borumensis* is usually coarsely scabrid towards the apex, while in *A. bipartita* it is glabrous and smooth.

Subsp. **contracta** (de Winter) Melderis in Bol. Soc. Brot., Sér. 2, **44**: 287 (1970). Type from S. Africa (Transvaal).
   Aristida scabrivalvis var. *contracta* de Winter in Kirkia, **3**: 132 (1963); Bothalia, **8**: 257 (1965).—W. D. Clayton, tom. cit.: 148 (1970). Type as above.

Botswana. N: Tsau-Nokaneng road, 1010 m., 20.iii.1961, *Vesey-FitzGerald 3282* (BM; SRGH). **Zambia.** S: lip of 5th Gorge, Victoria Falls Trust Area, c. 874 m., 23.ii.1963, *Bainbridge 756* (BM). **Rhodesia.** W: Wankie, Victoria Falls, 910 m., 4.iv.1956, *Robinson 1432* (SRGH). E: Melsetter, Sabi Valley 8 km. S. of Hot Springs, 520 m., 23.iv.1969, *Plowes 3185* (BM; SRGH).
   Also in Natal, the Transvaal and Tanzania. Often on heavy basalt soils in dry open situations in grassland, scrub and " mopane " savanna and along roadsides.

16. **Aristida bipartita** (Nees) Trin. & Rupr., Sp. Gram. Stip.: 144 (1842).—Dur. & Schinz, Consp. Fl. Afr. **5**: 801 (1894).—Wood, Natal Plants, **5**: pl. 483 (1908).— Henrard, Crit. Rev. Aristida **1** in Meded. Rijks-Herb. **54**: 54 (1926); Monogr. Aristida **2** in Meded. Rijks-Herb. **58a**: 194, t. 89 (1932).—Schweickerdt in Bothalia, **4**: 136 (1941).—Chippindall in Meredith, Grasses & Pastures of S. Afr.: 308, map 40 (1955).—Pilg. in Engl. & Prantl, Nat. Pflanzenfam. ed. 2, **14d**: 121 (1956).—de Winter in Bothalia, **8**: 257, fig. 24, 30, 159/9 (1965). Type from S. Africa.
   Chaetaria bipartita Nees, Fl. Afr. Austr.: 187 (1841). Type as above.

Perennial or subperennial, up to 65 cm. high, densely caespitose, with a short erect or oblique rhizome. Culms erect or ascending, usually slightly compressed, simple, glabrous or slightly pubescent, 1–3-noded; nodes glabrous. Leaf-sheaths firm, compressed, glabrous or minutely puberulous. Ligule a densely ciliolate rim; auricles long-barbate; collar glabrous or scaberulous. Leaf-laminae up to 20 × 0·2 cm., rigid, curved and conduplicate, scabrid above, glabrous, smooth or scaberulous beneath, with prominent whitish midrib and marginal nerves. Panicle large, usually up to 25 × 6–13 cm., effuse, with an angular scabrid axis; branches 2–10 cm. long, solitary (the lower mostly 2-partite), flexuous, spreading, scaberulous, naked for most of their length and bearing diffuse clusters of 1–4 spikelets at their ends, with glandular pubescent patches in the axils. Spikelets greenish or purplish. Glumes unequal, linear lanceolate, 1-nerved, usually glabrous, smooth, keeled, smooth or scaberulous on the keel, mucronate or shortly awned at the apex; the inferior up to 6·5–8 mm. long; the superior up to 6–7 mm. sometimes slightly emarginate. Lemma 5–6 mm. long, included in the glumes, glabrous, smooth, or scaberulous towards the apex, often purple-mottled; callus 0·4–0·5 mm. long, obtusely rounded, shortly barbate; beak or column absent, awns subequal, c. 7–14 mm. long; articulation absent.

Mozambique. LM: Sábiè, Moamba, i.1949, *Pimenta 52556* (LISC).
   Also in the Cape Prov., Orange Free State, the Transvaal, Natal and Lesotho. In open grassland.

This perennial or subperennial sp. is characterized by its very lax panicle with long spreading branches naked in their lower part, bearing a few spikelets on their ends, and by the inferior glume exceeding the superior in length. In the appearance of the panicle it resembles *A. scabrivalvis* subsp. *borumensis*, which is an annual grass with smaller spikelets and weaker development of the stereome strands of the leaf-laminae. In addition, in *A. scabrivalvis* subsp. *borumensis* the glumes are scabrid or hirtellous, and subequal or unequal (the inferior shorter than the superior); its lemma has usually coarse scabridules in rows towards the apex.

17. **Aristida wildii** Melderis in Bol. Soc. Brot., Sér. 2, **44**: 283, t. II, IV, fig. B (1970). Type: Botswana, N., 10 km. N. of Aha Hills, 12.iii.1965, *Wild & Drummond 6948* (BM, holotype; K, SRGH, isotypes).

Annual, 35–55 cm. high, delicate. Culms erect, slender, simple or branched, glabrous, smooth, 3–4-noded; nodes glabrous. Leaf-sheath glabrous, slightly keeled, tinged with purple towards the base. Ligule a short ciliate rim; auricles short-ciliate; collar glabrous. Leaf-laminae up to 22 × 0·1–0·15 cm., convolute, minutely hirtellous above, glabrous, smooth beneath. Panicle 20–22 cm. long, lax, effuse; axis slightly scaberulous; branches up to 6 cm. long (excluding spikelets), simple or 2-nate, naked for most of their length, scaberulous, with a dark coloured glandular patch in the axils, ± spreading or the lower ones sometimes reflexed, bearing 1–3 spikelets towards the ends. Spikelets pallid or tinged with purple. Glumes unequal, the lower longer than the upper, hyaline-membranous, 1-nerved, keeled, slightly scaberulous on the keel, otherwise glabrous; the inferior 7–9 mm. long, acute; the superior 6–7·5 mm. long, erose and obtuse at the apex. Lemma 10·5–11 mm. long, exserted from the glumes, keeled, often mottled with purple, glabrous in the lower part, scaberulous on the keel and lateral nerves towards the apex; callus 0·5 mm. long, short-barbate, with a minute, obtuse, glabrous tip; column absent; awns unequal, scabrous, the central 1·1–1·6 cm. long, the lateral ones 0·6–1·3 cm. long; articulation absent.

Botswana. N: 10 km. N. of Aha Hills, 12.iii.1965, *Wild & Drummond* 6948 (BM, holotype; K, SRGH, isotypes).

Not known from elsewhere. In *Combretum apiculatum-Terminalia prunioides* savanna on shallow sand overlying frequently outcropping limestone.

This species is closely related to *A. effusa* Henrard, occurring in SW. Africa, from which it differs in having the inferior glume distinctly longer than the superior, a lemma glabrous in the lower part and scaberulous on the keel, and lateral nerves towards the apex. In the appearance of the panicle *A. wildii* resembles *A. bipartita*, from which it can be distinguished by its annual habit and by a lemma conspicuously exserted from the glumes.

18. **Aristida rhiniochloa** Hochst. in Flora, **38**: 200 (1855).—Dur. & Schinz, Consp. Fl. Afr. **5**: 808 (1894).—Engl., Pflanzenw. Ost-Afr. **C**: 107 (1895).—Rendle, Cat. Afr. Pl. Welw. **2**: 204 (1899).—Dinter in Fedde, Repert. **15**: 342 (1918), (" *rhinochloa* ").—Henrard, Crit. Rev. Aristida 3 in Meded. Rijks-Herb. **54b**: 510 (1928); Monogr. Aristida **2** in Meded. Rijks-Herb. **58a**: 242, t. 115 (1929).—Stent & Rattray in Proc. Rhod. Sci. Ass. **32**: 46 (1933), sub *A. rhiniochloa* forma *rigidiseta*, " nom. nud.".—Schweickerdt in Bothalia, **4**: 146 (1941).—Sturgeon in Rhod. Agric. Journ. **51**: 507 (1954).—Chippindall in Meredith, Grasses & Pastures of S. Afr.: 310 (1955).—Pilg. in Engl. & Prantl, Nat. Pflanzenfam. ed. 2, **14d**: 121 (1956).—de Winter in Bothalia, **8**: 249, fig. 18, 19, 20, 159/4 (1965).—W. D. Clayton in F.T.E.A., Gramineae: 147 (1970).—Launert in Merxm., Prodr. Fl. SW. Afr. **160**: 33 (1970). Type from Ethiopia.
  Aristida rigidiseta Pilg. in Engl., Bot. Jahrb. **51**: 413 (1914).—Henrard, Crit. Rev. Aristida **3**, op. cit. **54b**: 516 (1928).—Sturgeon, loc. cit. Type from SW. Africa.
  Aristida andoniensis Henrard, Crit. Rev. Aristida 3, op. cit. **54b**: 691 (1928); Monogr. Aristida **2** in Meded. Rijks-Herb. **58a**: 243 (1932). Type from SW. Africa.

Annual, up to 65 cm. high, caespitose, very scabrid and coarse. Culms erect or geniculately ascending, branched in the lower part and from the middle nodes, retrorsely scabrous, 3–many-noded; nodes minutely pubescent. Leaf-sheaths laxly embracing the culm, keeled, scabrous. Ligule long-ciliate; auricles long-barbate; collar minutely pubescent or glabrous. Leaf-laminae 10–20 × 0·2–0·4 cm., flat, glaucous, very scabrous on both surfaces. Panicle up to 20 cm. long, exserted, effuse or contracted and much interrupted, with very scabrid axis and branches; branches 2-nate or solitary, with dark-coloured, shortly pubescent glandular patches in the axils. Spikelets coarse, reddish-green with a black spot at the base. Glumes broadly lanceolate, shortly pubescent on the back, scaberulous on the keel, 1-nerved, awned; the inferior up to 17 mm. long, gradually tapering into a short awn up to 1·5 mm. long, the superior up to 15 mm. long, usually with 2 more or less acute lobes at the apex and a short awn 0·5–1 mm. long from the sinus. Lemma usually c. 11 mm. long, prominently nerved, with coarse antrorsely curved scabridules on the nerves, very rarely almost glabrous, keeled on the back, deeply grooved ventrally, with inrolled margins; callus 0·5–0·75 mm. long, subobtuse, barbate; column absent; awns subequal, usually 18–30 mm. long, rigid, erect or spreading, very scabrous, triquetrous, sometimes slightly winged at the base; articulation absent.

Caprivi Strip. Mpila I., c. 910 m., 13.i.1959, *Killick & Leistner* 3367 (SRGH). **Botswana.** N: Near Pandamatenga, 28.iii.1961, *Vesey-FitzGerald* 3361 (SRGH). **Zambia.** E: Luangwa Valley, Chilongozi Reserve, 14.iv.1963, *Verboom* 947 (BM; K).

S: Lutiri R., Zambezi Valley, Lake Kariba, 760 m., 5.ix.1959, *Vesey-FitzGerald* 2626 (SRGH). **Rhodesia.** W: Shangani, Gwampa Forest Reserve, ii.1955, *Goldsmith* 116 (K; SRGH). C: Sebakwe R. near Que Que, 6.iii.1961, *Vesey-FitzGerald* 3102 (SRGH). E: Melsetter, 910 m., 16.ii.1954, *Crook* 526 (SRGH). S: Sabi Valley near Devuli R., 27.i–2.ii.1948, *Rattray* 1299 (SRGH).

Also in SW. Africa, from Mauritania to Eritrea, Tanzania and the Transvaal. On dry sandy and basalt soils, sandy-clay loams and stony ground, in disturbed areas, along roadsides, on gravelly flats, on rocky hill slopes, under trees in valleys, in scrub, on kopjies, at edges of shallow depressions.

A distinct sp. which can be easily recognized by its annual habit, retrorsely scabrid culms, scabrous leaves, a large panicle with scabrid axis and branches, broadly lanceolate glumes, of which the inferior is longer than the superior, a long lemma (c. 11 mm.), with rows of coarse antrorse scabridules from the base to the apex; by the absence of a column and by rigid triquetrous awns, more or less winged at the base.

According to de Winter, *A. rhiniochloa* seems to be related to *A. hordeacea*, because both have a similar leaf anatomy (expanded leaf-laminae with abaxially protruding midribs and a very similar type of vascular bundle units). In addition, they are annuals characterized by having scabrid ventrally grooved lemmas without a column at the apex. They can, however, be easily distinguished by the type of the inflorescence and by the presence or absence of the articulation at the apex of the lemma. In *A. rhiniochloa* the panicle is effuse or contracted and much interrupted and the lemma has no articulation, while in *A. hordeacea* the panicle is very dense, spiciform, and the lemma articulated.

19. **Aristida hordeacea** Kunth, Rév. Gram. **2**: 517, t. 173 (1831).—Dur. & Schinz, Consp. Fl. Afr. **5**: 803 (1894).—Rendle, Cat. Afr. Pl. Welw. **2**, 1: 204 (1899).— Dinter in Fedde, Repert. **15**: 342 (1918), (" *hordacea* ").—Henrard, Crit. Rev. Aristida 2 in Meded. Rijks-Herb. **54a**: 241 (1927); Monogr. Aristida 1 in Meded. Rijks-Herb. **58**: 140, t. 54 (1929).—Schweickerdt in Bothalia, **4**: 172 (1941).— Sturgeon in Rhod. Agric. Journ. **51**: 503 (1954).—Chippindall in Meredith, Grasses & Pastures of S. Afr.: 317 (1955).—Pilg. in Engl. & Prantl, Nat. Pflanzenfam. ed. 2, **14d**: 122 (1956).—Jackson & Wiehe, Annot. Check List Nyasal. Grass.: 29 (1958). —de Winter in Bothalia, **8**: 244 fig. 13, 14, 55, 159/2 (1965).—W. D. Clayton, F.T.E.A., Gramineae: 150, fig. 48 (1970).—Launert in Merxm., Prodr. Fl. SW. Afr. **160**: 31 (1970). TAB. **31**. Type from Senegambia.

Aristida steudeliana Trin. & Rupr., Sp. Gram. Stip.: 155 (1842). Type from Ethiopia.

Aristida hordeacea var. *longiaristata* Henrard, Crit. Rev. Aristida 2 in Meded. Rijks-Herb. **54a**: 244 (1927). Type from SW. Africa.

Aristida pseudohordeacea Stent & Rattray in Proc. Rhod. Sci. Ass. **32**: 45 (1933). Type: Rhodesia, Wankie, Victoria Falls, 900 m., 12.iii.1932, *Brain* GH. 5827 (SRGH, holotype).

Annual, 10–90 cm. high, caespitose. Culms erect or ascending, branched from the base and lower nodes, compressed, densely pubescent, sometimes becoming more or less glabrous, 3–4-noded; nodes pubescent. Leaf-sheaths keeled, pubescent. Ligule shortly ciliate; auricles shortly barbate; collar glabrous. Leaf-laminae c. 30 × 1 cm., more or less glaucous, flat or folded, scabrous on both surfaces or hirtellous above, becoming glabrous beneath. Panicle 25 × 1–3 cm., exserted, very dense, linear-oblong or subovate, spiciform or interrupted at the base, with a densely pubescent axis; branches solitary but much divided from the base. Spikelets linear-lanceolate, greenish or pallid, fascicled; pedicels pubescent. Glumes lanceolate, keeled, shortly and densely pubescent, awned or 2-fid at the apex with an awn from the sinus; the inferior 5–9 mm. long, with an awn 2·5–5 mm. long; the superior 6–10 mm. long, with an awn 1–4 mm. long. Lemma 5–7·5 mm. long, narrowly linear, fusiform, with many longitudinal rows of spiny hairs, ventrally furrowed, with a distinct articulation between the apex of the lemma and the base of the awns; column absent; callus c. 0·5 mm. long, obtuse, densely barbate; awns 15–50 mm. long, subequal, erect, scabrous. $2n = 22$.

Botswana. N: L. Ngami near Sehitwa, 24.iii.1961, *Vesey-FitzGerald* 3330 (K; SRGH). SE: Serowe, Kalamabele, 980 m., 26.iii.1957, *de Beer* 54 (LISC; SRGH). **Zambia.** N: Mpika Distr., Kapiri Plots 45·1 km. NW. of Mfuwe, c. 700 m., 21.iii.1969, *Astle* 5632 (BM; SRGH). S: Victoria Falls, Livingstone, 29.iii.1961, *Vesey-FitzGerald* 3366 (BM; SRGH). **Rhodesia.** N: Mtoko, Suskwe area, 16.iii.1953, *Phelps* 35 (SRGH). W: Nymandhlovu, Pasture Station, 5.iii.1953, *Plowes* 1552 (SRGH). C: Gatooma, 1142 m., 20.iii.1930, *Brain* 939 (BM; SRGH). S: Ndanga, 270 m., 1.iv.1961, *Phipps* 2917 (BM; SRGH). **Malawi.** N: Nyika Plateau, ii–iii.1903, *McClounie* sn. (K). **Mozambique.** N: Eráti, Namapa, near CICA Experimental

Tab. 31. ARISTIDA HORDEACEA. 1, habit (×½); 2, junction of leaf-sheath and lamina showing ligule (×3); 3, spikelet (×4½); 4, glumes (×8); 5, lemma (×8); 6, palea (×12); 7, sexual organs, with lodicules detached (×12), all from *Harrington* 118. From F.T.E.A.

Station, base of Serra M'Poge, 23.iii.1960, *Lemos & Macuácua* 57 (BM; COI; LISC; LMA; SRGH).

Also in SW. Africa, Angola, Uganda, Kenya, Tanzania; widely distributed throughout tropical Africa. Grows on sandy, clayey and black basaltic soils, in open grasslands and woodlands mainly in shallow depressions, on edges of small pans, banks of streams and mopane velds.

This annual sp. seems to be related to *A. rhiniochloa* (see page 115) and is readily distinguishable from all the others by compressed densely pubescent culms, flat usually glaucous leaf-laminae, subovate spike-like panicles, shortly and densely compact pubescent glumes, a fusiform lemma with many longitudinal rows of spiny hairs and by an articulation between the apex of the lemma and the base of the awns.

20. **Aristida diffusa** Trin. in Mém. Acad. Imp. Sci. Petersb., Sér 6, **1**: 86 (1830).—Kunth, Enum. Pl. **1**: 193 (1833).—Henrard, Crit. Rev. Aristida **1** in Meded. Rijks-Herb. **54**: 142 (1926); Monogr. Aristida **1** in Meded. Rijks-Herb. **58**: 96, t. 28 (1929).— de Winter in Bothalia, **8**: 275, fig. 53, 54, 55, 159/18 (1965). Type from S. Africa (Cape).

Aristida hystrix Thunb., Prodr. Pl. Cap. **1**: 19 (1794) non L.f. (1781); Fl. Cap. **1**: 394 (1813).—Henrard, Crit. Rev. Aristida **2** in Meded. Rijks-Herb. **54a**: 252 (1927). Type from S. Africa (Cape).

Aristida vestita var. *diffusa* (Trin.) Trin. & Rupr., Sp. Gram Stip.: 157 (1842). Type as for *Aristida diffusa*.

Aristida vestita var. *brevistipitata* Trin. & Rupr., tom. cit.: 158 (1842), "*brevistipitata*". —Henrard, Crit. Rev. Aristida **3** in Meded. Rijks-Herb. **54b**: 665 (1928). Type from S. Africa (Cape).

Aristida vestita var. *densa* Trin. & Rupr., loc. cit.—Henrard, Crit. Rev. Aristida **3**, loc. cit. Type from S. Africa (Cape).

Aristida vestita var. *eckloniana* Trin. & Rupr., loc. cit.—Henrard, tom. cit.: 667 (1928). Type from S. Africa (Cape).

Aristida vestita var. *schraderana* Trin. & Rupr., loc. cit. ("*schraderiana*").— Henrard, tom. cit.: 668 (1928). Type from S. Africa (Cape).

Aristida vestita var. *pseudohystrix* Trin. & Rupr., loc. cit. Type from S. Africa (Cape).

Aristida pseudohystrix (Trin. & Rupr.) Steud., Syn. Pl. Glum.: 142 (1854).— Henrard, Crit. Rev. Aristida **3** in Meded. Rijks-Herb. **54b**: 470 (1928). Type as for *Aristida vestita* var. *pseudohystrix*.

Aristida diffusa var. *brevistipitata* (Trin. & Rupr.) Henrard, Crit. Rev. Aristida **3** in tom. cit.: 665 (1928). Type as for *Aristida vestita* var. *brevistipitata*.

Aristida diffusa var. *densa* (Trin. & Rupr.) Henrard, loc. cit.; Monogr. Aristida **1** in Meded. Rijks-Herb. **58**: 97 (1929). Type as for *Aristida vestita* var. *densa*.

Aristida diffusa var. *eckloniana* (Trin. & Rupr.) Henrard, tom. cit.: 667 (1928); Monogr. Aristida **1**, tom. cit.: 97, t. 29 (1929). Type as for *Aristida vestita* var. *eckloniana*.

Aristida diffusa var. *pseudohystrix* (Trin. & Rupr.) Henrard, Crit. Rev. Aristida **3**, tom. cit.: 471 (1928); Monogr. Aristida **1**, loc. cit.—Schweickerdt in Bothalia, **4**: 169 (1941). Type as for *Aristida vestita* var. *pseudohystrix*.

Aristida diffusa var. *schraderana* (Trin. & Rupr.) Henrard, Crit. Rev. Aristida **3**, tom. cit.: 668 (1928); Monogr. Aristida **1**, loc. cit. Type as for *Aristida vestita* var. *schraderana*.

Aristida diffusa var. *genuina* Henrard, Monogr. Aristida **1**, loc. cit.—Schweickerdt in Bothalia, **4**: 166 (1941). Type as for *Aristida diffusa*.

Perennial, up to 75 cm. high, densely to laxly caespitose. Culms erect, slender, simple or rarely branched, glabrous, 1- to several-noded; nodes glabrous. Leaf-sheaths firm, glabrous, or the lower densely woolly. Ligule a short-ciliate rim; auricles pubescent or glabrous; collar glabrous. Leaf-laminae up to 30 × 0·25–3 cm., involute, setaceous, curved or flexuous, the upper exceeding the panicle, scaberulous or hispidulous above, smooth beneath. Panicle up to 30 cm. long, or longer, lax, open, with a smooth axis; branches c. 15 cm. long, 2- or 3-nate, filiform or capillary, more or less spreading, remote, flexuous, scaberulous or smooth. Spikelets yellowish, yellowish-brown to purple, purplish-brown, suberect to nodding. Glumes unequal, rather firm, 1-nerved, glabrous, obtuse or slightly emarginate with a glabrous keel; the inferior 4–9 mm. long, fairly broad, obtuse; the superior 10–17 mm. long, narrow, with a subacute but slightly 2-fid apex. Lemma 9–13 mm. long, smooth, or slightly scaberulous towards the apex; callus c. 1 mm. long, 2-fid, barbate; column 2–6 mm. long, slightly twisted; awns fine, subequal, scabrous, the central awn up to 23–25 mm. long, the later ones

20–30 mm. long; articulation between the apex of the lemma and the base of the column.

Subsp. **burkei** (Stapf) Melderis in Bol. Soc. Brot., Sér, 2, **44** : 287 (1970). Type from S. Africa (Orange Free State).

Aristida burkei Stapf in Harv. & Sond., F.C. **7**: 557 (1899).—Henrard, Crit. Rev. Aristida **1** in Meded. Rijks-Herb. **54**: 64 (1926); Monogr. Aristida **2** in Meded. Rijks-Herb. **58a**: 184, t. 81 (1932). Type as above.

Aristida diffusa var. *burkei* (Stapf) Schweickerdt in Notizbl. Bot. Gart. Berl. **14**: 195 (1938); in Bothalia, **4**: 167 (1941).—Chippindall in Meredith, Grasses & Pastures of S. Afr.: 316, pl. 8, map 45 (1955).—de Winter in Bothalia, **8**: 277, fig. 55, 56 (1965). Type as above.

Culms usually 3- to 5-noded. Panicle up to 30 cm. long, lax and open. Spikelets yellowish or yellow-brown. Inferior glume c. 6 mm. long, with obtusely emarginate apex; the superior usually c. 12 mm. long, with a subacute slightly 2-fid apex. Lemma usually less than 11 mm. long (excluding the callus). $2n = 22$.

Rhodesia. N: Lomagundi, Umvukwes Range, 1460 m., 25.ii.1959, *Phipps, Drummond & Jackson* 1522 (BM; COI; SRGH).
Also in the Cape Prov., Lesotho, Orange Free State, Natal, the Transvaal. On serpentine and rubbly soils in hilly grasslands.

Subsp. *burkei* is widespread in S. Africa extending also into Rhodesia, whilst subsp. *diffusa* is confined to the southern Cape Prov. In some areas common to both subspp. intermediates between them are met with. Subsp. *diffusa* differs from subsp. *burkei* in having 1–2-noded culms and a shorter (up to 15 cm. long), often somewhat contracted panicle. Its spikelets are purple, purplish-brown or brown; the inferior glume is 4–9 mm. long, obtuse; the superior 13–17 mm. long, with an obtuse 2-dentate apex; the lemma usually c. 11–13 mm. long (excluding the callus).

21. **Aristida vestita** Thunb., Prodr. Pl. Cap.: 19 (1794); F.C. **1**: 394 (1813).—Hack. in Engl., Bot. Jahrb. **11**: 400 (1889).—Dur. & Schinz, Consp. Fl. Afr. **5**: 810 (1894), pro parte.—Engl., Pflanzenw. Ost-Afr. C: 107 (1895).—Rendle in Cat. Afr. Pl. Welw. **2**: 204 (1899).—Stapf in Harv. & Sond., F.C. **7**: 561 (1899), pro parte. Wood, Natal Pl. **5**: pl. 404 (1904).—Monro in Proc. Rhod. Sci. Ass. **6**: 37 (1906). —Henrard, Crit. Rev. Aristida **3** in Meded. Rijks-Herb. **54b**: 663 (1928); Monogr. Aristida **1** in Meded. Rijks-Herb. **58**: 94, t. 27 (1929).—Stent & Rattray in Proc. Rhod. Sci. Ass. **32**: 44 (1933).—Schweickerdt in Bothalia, **4**: 162 (1941).— Sturgeon in Rhod. Agric. Journ. **51**: 500 (1954).—Chippindall in Meredith, Grasses & Pastures of S. Afr.: 315 (1955).—Pilg. in Engl. & Prantl, Nat. Pflanzenfam. ed. 2, **14d**: 123 (1956).—de Winter in Bothalia, **8**: 279, fig. 36, 57, 159/19 (1965).—W. D. Clayton, F.T.E.A., Gramineae: 152 (1970).—Launert in Merxm., Prodr. Fl. SW. Afr. **160**: 34 (1970). Type from S. Africa (Cape).

Chaetaria vestita (Thunb.) Beauv., Agrost.: 30 (1812).—Roem. & Schult. in L., Syst. Veg., ed. nov., **2**: 392 (1817). Type as above.

Aristida lanuginosa Burch., Trav. Int. S. Afr., **2**: 226 (1824).—Henrard, Crit. Rev. Aristida **2** in Meded. Rijks-Herb. **54a**: 287 (1927). Type from S. Africa (Griqualand West).

Arthratherum vestitum (Thunb.) Nees, Fl. Afr. Austr.: 174 (1841), pro parte. Type as for *Aristida vestita*.

Aristida vestita forma *amplior* Hack. in Engl., Bot. Jahrb. **11**: 400 (1889).— Henrard, Crit. Rev. Aristida **3** in Meded. Rijks-Herb. **54b**: 664 (1928). Type from S. Africa (Griqualand West).

Aristida flocciculmis Mez in Fedde, Repert. **17**: 147 (1921).—Henrard, Crit. Rev. Aristida **1** in Meded. Rijks-Herb. **54**: 182 (1926). Type from S. Africa (Cape).

Perennial, up to 85 cm. high, robust, densely caespitose. Culms rigid, simple, sometimes branched from the base or from the nodes, pubescent or woolly in the lower part, often glabrous upwards, 3–4-noded; nodes glabrous. Lower leaf-sheaths broad and papery, straw-coloured, lanate on the back and along the margins, soon becoming glabrous. Ligule a very shortly ciliate rim; auricles with a tuft of woolly hairs. Leaf-laminae up to 24×0.4 cm., flat at the base, involute in the upper part, shortly pubescent or somewhat villous above, glabrous beneath. Panicle up to 20×12 cm., pyramidal, effuse, open or laxly contracted; axis glabrous or with scattered white hairs, sometimes barbate in the lower axils, with fine, suberect or somewhat spreading scabrous branches. Spikelets greenish or brownish. Glumes unequal, firm, 1-nerved, glabrous; the inferior 4·5–7 mm. long, lanceolate, rounded at the apex; the superior 9–13 mm. long, narrowly lanceolate, somewhat

denticulate along the margins towards the apex, slightly emarginate. Lemma
7–11 mm. long, glabrous, often mottled with purple; callus 1–1·5 mm. long, 2-fid,
barbate; column 2–7 mm. long, strongly twisted; awns unequal; the central
20–35 mm. long, the lateral ones shorter; articulation between the apex of the
lemma and the base of the column.

Zambia. S: near Sinazongwe, 18.iv.1965, *Astle* 3078a (K; SRGH). **Rhodesia.** N:
Darwin, Chimanda Reserve, 18.iii.1959, *Cleghorn* 466 (BM; SRGH). W: Shangani/Bubi,
Gwampa Forest Reserve, c. 910 m., vi.1956, *Goldsmith* 138 (BM; COI; SRGH). C:
Que Que, Sebakwe R., 6.iii.1961, *Vesey-FitzGerald* 3096 (K; SRGH). E: Umtali, Carolina
B. Farm near Pounsley, 1040 m., 27.ii.1957, *Phipps* 545 (SRGH). S: Sabi Valley,
Devuli Ranch, 21.i.1948, *Fisher* 1402 (SRGH). **Mozambique.** T: 148 km. on the
railway from Tete, 18.v.1948, *Mendonça* 4321 (LISC).
Also in the Cape Prov., SW. Africa, Orange Free State, the Transvaal and Tanzania.
On limestone, black clayey and veld soils.

Differs from *A. mollissima*, which also has woolly culms, in having shorter glumes,
a 2-fid callus and a shorter column of the awns.

22. **Aristida meridionalis** (Stapf) Henrard, Crit. Rev. Aristida 2 in Meded. Rijks-Herb.
    54a: 344 (1927); Monogr. Aristida 1 in Meded. Rijks-Herb. **58:** 95, t. 27 (1929).—
    Stent & Rattray in Proc. Rhod. Sci. Ass. **32:** 44 (1933).—Schweickerdt in Bothalia,
    4: 163 (1941).—Sturgeon in Rhod. Agric. Journ. **51:** 500 (1954).—Chippindall in
    Meredith, Grasses & Pastures of S. Afr.: 315 (1955).—Pilg. in Engl. & Prantl., Nat.
    Pflanzenfam., ed. 2, **14d:** 123 (1956).—de Winter in Bothalia **8:** 284, fig. 62, 63, 199/
    22 (1965).—W. D. Clayton, F.T.E.A., Gramineae: 153 (1970).—Launert in Merxm.,
    Prodr. Fl. SW. Afr. **160:** 32 (1970). TAB. **32.** Type from SW. Africa.
    Aristida stipoides var. *meridionalis* Stapf in Harv. & Sond. F.C. **7:** 562 (1899).
    Type from S. Africa (Griqualand West).

Perennial, up to 2 m. high, very robust, densely caespitose. Culms erect, simple
or slightly branched, glabrous, 2–3-noded; nodes often dark-coloured, glabrous.
Lower leaf-sheath glabrous or woolly, afterwards becoming glabrous, the upper
ones glabrous, smooth, somewhat loose. Ligule a short-ciliate rim; auricles
barbate, with a tuft of woolly hairs; collar glabrous. Leaf-laminae up to 50 × 0·5
cm., involute, scabrous above, glabrous, smooth beneath. Panicle up to 80 cm.
long and over 20 cm. wide, effuse, very lax, often nodding, many-flowered, with
smooth axis; branches 2–3-nate with filiform branchlets. Spikelets yellowish,
yellowish-brown or tinged with purple. Glumes unequal, 1-nerved, glabrous;
the inferior 5–7 mm. long, obtuse or emarginate, slightly ciliate upwards along the
margins; the superior 10–15 mm. long, 2-fid at the apex. Lemma 7–9 mm. long,
mottled with purple, smooth or slightly punctulate in the lower part, scaberulous
towards the tip; callus 1·5–2 mm. long, barbate, distinctly 2-fid; column 5–20 mm.
long, slender, strongly twisted, scabrous or nearly glabrous; awns up to 5 cm.
long, subequal; articulation between the apex of the lemma and the base of the
column. 2n = 22.

Caprivi Strip. Near Linyati, 910 m., 28.xii.1958, *Killick & Leistner* 3163 (K; SRGH).
Botswana. N: Bushmans Pits, 96 km. E. of Maun, 27.iii.1961, *Vesey-FitzGerald*
3341 (BM; SRGH). SW: Ghanzi, NW. Kalahari Crownland, 995 m., ii.1952,
de Beer D61 (K; SRGH). SE: Bakhatla Reserve, 12.iv.1931, *Pole Evans* 3157 (K;
SRGH). **Zambia.** B: 3·2 km. S. of Shangombo, 1020 m., 9.viii.1952, *Codd* 7478 (K;
PRE). N: Mporokoso, 8 km. N. of Muzombwe, W. side of Mweru-wa-Ntipa, 1070 m.,
16.iv.1961, *Phipps & Vesey-FitzGerald* 3223 (BM; SRGH). S: Mazabuka, iii.1932,
Trapnell 1067 (K; SRGH). **Rhodesia.** N: Darwin, Chimanda Reserve, 18.iii.1959,
Cleghorn 473 (BM; SRGH). W: Plumtree, 1328 m., 8.iii.1930, *Brain* 789 (K; SRGH).
C: Castle Combe near Marandellas, 1445 m., vi.1959, *Wicken* 557 (SRGH). E:
Chipinga, 1220 m., iv.1934, *Brain* 10676 (K; SRGH). S: Fort Victoria, Makoholi
Experimental Station, 18.ii.1948, *Newton & Juliasi* 32 (K; SRGH). **Mozambique.**
SS: Caniçado, Região de Massingir, 22.v.1948, *Torre* sn. (LISC).
Also in the Cape Prov., Orange Free State, Transvaal, SW. Africa, Angola and Tanzania.
Mainly on sandy soils in open situations, in open woodland, scrub, vlei, on edges of water-
logged depressions, on sand dunes (Kamba Bay, Lake Tanganyika).

This sp. is easily recognizable by a large very lax and effuse panicle and by the presence
of a tuft of woolly hairs at the mouth of the leaf-sheath, and by a 2-fid callus of the lemma.
It is closely allied to *A. stipoides* from which it differs by the perennial habit and a shorter
column of the lemma.

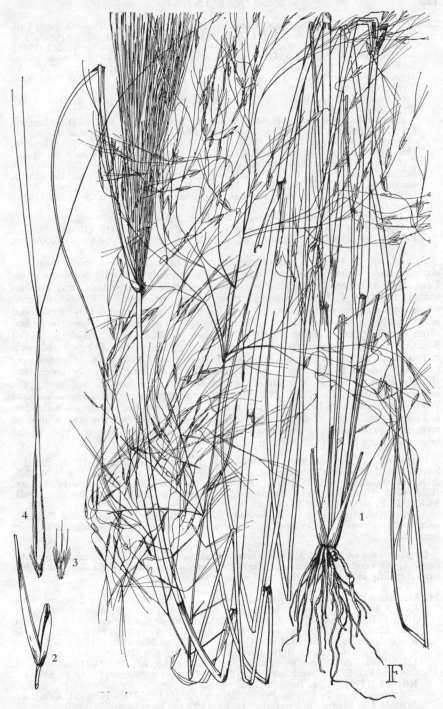

Tab. 32. ARISTIDA MERIDIONALIS. 1, habit (× ½); 2, glumes (× 6); 3, base of lemma showing 2-fid callus (× 6); 4, lemma (× 6), all from *Torre* 7228.

23. **Aristida stipoides** Lam., Tabl. Encycl. Méth. Bot. **1**: 157 (1791).—Henrard, Crit. Rev. Aristida **3** in Meded. Rijks-Herb. **54b**: 591 (1928); Monogr. Aristida **1** in Meded. Rijks-Herb. **58**: 93, t. 26 (1929).—Schweickerdt in Bothalia, **4**: 161 (1941). —Chippindall in Meredith, Grasses & Pastures of S. Afr.: 315 (1955).—de Winter in Bothalia, **8**: 286, fig. 21, 64, 65, 159/23 (1965).—W. D. Clayton, F.T.E.A., Gramineae: 153 (1970).—Launert in Merxm., Prodr. Fl. SW. Afr. **160**: 34 (1970). Type from Senegal.

Chaetaria lamarkii Roem. & Schult. in L., Syst. Veg. ed. nov., **2**: 393 (1817). Type from Senegal.

Aristida lamarkii (Roem. & Schult.) Steud., Nom. Bot. **1**: 69 (1821).—Henrard, Crit. Rev. Aristida **2** in Meded. Rijks-Herb. **54a**: 282 (1927). Type as above.

Aristida amplissima Trin. & Rupr., Sp. Gram. Stip.: 155 (1842).—Henrard, Crit. Rev. Aristida **1** in Meded. Rijks-Herb. **54**: 24 (1926). Type from Senegal.

Aristida gracillima Oliv. in Trans. Linn. Soc. **29**: 173, t. 114 (1875).—Dur. & Schinz, Consp. Fl. Afr. **5**: 803 (1894).—Engl., Pflanzenw. Ost-Afr. **C**: 107 (1895); op. cit. **A**: 14, 17, 21, 51 (1895).—Hack. in Bull. Herb. Boiss. **4**, Append. **3**: 18 (1896).—Henrard, Crit. Rev. Aristida **1** in Meded. Rijks-Herb. **54**: 212 (1926). Type from E. Africa.

Aristida fontismagni Schweickerdt in Engl., Bot. Jahrb. **76**: 220 (1954).— Chippindall in Meredith, Grasses & Pastures of S. Afr.: 315 (1955).—Pilg. in Engl. & Prantl, Nat. Pflanzenfam. ed. 2, **14d**: 123 (1956). Type from SW. Africa.

Annual, up to 1·5 m. high, robust, laxly caespitose. Culms erect, simple or slightly branched from the upper nodes, glabrous, usually 3-noded; nodes glabrous. Leaf-sheaths slightly scaberulous or glabrous. Ligule a short-ciliate rim; auricles with a tuft of woolly hairs; collar glabrous. Leaf-laminae up to 30 × 0·3–0·5 cm., when expanded, involute or more or less flat at the base, prominently nerved and scabrous above, smooth beneath. Panicle up to 50 cm. long, effuse and very lax, often nodding, often few-flowered; branches glabrous or scaberulous, 2- or 3-nate with filiform branchlets; axis smooth. Spikelets pallid or tinged with red or purple, especially at the base, sometimes with a black spot. Glumes very unequal, 1-nerved, glabrous; the inferior 5–7 mm. long, broadly lanceolate, slightly 2-fid, minutely serrulate-ciliate at the margins towards the apex; the superior 15-20 mm. long, narrowly linear with a mucro from a 2-fid apex. Lemma 7–9 mm. long, smooth below, strongly punctulate towards the apex; callus slender, c. 2 mm. long, 2-fid, barbate; column 2–3 cm. long, prominently twisted, scabrous; awns subequal, scabrous, 3·5–5·5 cm. long; articulation between the apex of the lemma and the base of the column.

Botswana. N: Tsau, 980 m., 19.iii.1961, *Vesey-FitzGerald* 3263 (BM; SRGH). **Zambia.** B: Mongu Airport, 22.iii.1964, *Verboom* 1317 (BM). S: Machili, 20.ii.1961, *Fanshawe* 6281 (SRGH). **Rhodesia.** W: Wankie Game Reserve, 910 m., *Wild* 4793 (SRGH). S: Umzingwane, Irisvale Ranch, Balla Balla, iii.1961, M.R.S.H., P. 40 (BM; SRGH).

Also in SW. Africa and from Senegal, the French Sahara, Sudan and Ethiopia, southwards to Tanzania. On sandy and basaltic soils in grassland and scrub, also in peaty dambos.

In general appearance *A. stipoides* resembles *A. meridionalis* and similarly to the latter has a tuft of woolly hairs at the mouth of the leaf-sheaths and a 2-fid callus, but differs from it mainly in the annual habit and a longer column of the awns.

24. **Aristida mollissima** Pilg. in Engl., Bot. Jahrb. **40**: 80 (1907).—Henrard, Crit. Rev. Aristida **2** in Meded. Rijks-Herb. **54a**: 354 (1927); Monogr. Aristida **1** in Meded. Rijks-Herb. **58**: 99, t. 30 (1929).—Stent & Rattray in Proc. Rhod. Sci. Ass. **32**: 44 (1933).—Schweickerdt in Bothalia, **4**: 170 (1941).—Sturgeon in Rhod. Agric. Journ. **51**: 501 (1954).—Chippindall in Meredith, Grasses & Pastures of S. Afr.: 316 (1955).—Pilg. in Engl. & Prantl, Nat. Pflanzenfam. ed. 2, **14d**: 123 (1956).—de Winter in Bothalia, **8**: 287, fig. 66, 67, 159/24 (1965).—W. D. Clayton, F.T.E.A., Gramineae: 155 (1970).—Launert in Merxm., Prodr. Fl. SW. Afr. **160**: 32 (1970). Type from SW. Africa.

Aristida elymoides Mez in Fedde, Repert. **17**: 148 (1921).—Henrard, Crit. Rev. Aristida **1** in Meded. Rijks-Herb. **54**: 166 (1926). Type from SW. Africa.

Perennial, up to 90 cm. high, densely caespitose. Culms erect, robust, simple or rarely branched in the upper part, densely lanate-tomentose or densely woolly, especially towards the base, 3–7-noded; nodes glabrous, smooth. Leaf-sheaths lanate-tomentose to glabrous, usually tight. Ligule a shortly pubescent or ciliate rim; auricles with a tuft of woolly hairs; collar glabrous or minutely pubescent. Leaf-laminae up to 30 × 0·3 cm., involute, nearly filiform, scaberulous above, glabrous, smooth beneath. Panicle up to 20 cm. long, shortly exserted, narrow, very dense and almost spiciform; axis densely lanate; branches numerous, very short, appressed, scabrid. Spikelets nearly sessile, pallid or purplish. Glumes very unequal, narrowly lanceolate, usually 1-nerved, scabrid on the keel, finely scaberulous towards the apex, shortly awned; the inferior 8–10 mm. long; the superior 12·5–22 mm. long. Lemma 7–8·5 mm. long, minutely punctulate; callus 2–2·5 mm. long, very acute, elongate, short-barbate; column 16–27 mm. long, scabrous, twisted; awns 2·8–6 cm. long, subequal, scabrous, more or less spreading; articulation between the apex of the lemma and the base of the column.

Panicle up to 20 × 5 cm., spiciform, contracted, very dense -   -   subsp. *mollissima*
Panicle up to 30 cm. long, narrow, effuse, lax, more or less divaricately branched
                                                              subsp. *argentea*

Subsp. **mollissima**

Botswana. N: 69 km. W. of Nokaneng, 12.iii.1965, *Wild & Drummond* 6901 (BM; K; SRGH). **Rhodesia.** W: Shangani/Bubi, 910 m., ii.1956, *Goldsmith* 34 (SRGH). C: Mtao Forest Reserve, 28.ii.1948, *Robinson* 300 (SRGH).
Also in the Cape Prov., SW. Africa and Kenya. Occurs on sandy soils in dry grassland or savanna bushland.

This subsp. is closely allied to *A. stipitata* subsp. *stipitata*, but differs mainly in having densely lanate or woolly culms.

Subsp. **argentea** (Schweickerdt) Melderis in Bol. Soc. Brot., Sér. 2, **44**: 288 (1970). Type from S. Africa (Transvaal).
   Aristida argentea Schweickerdt in Engl., Bot. Jahrb. **76**: 218 (1954).—Sturgeon in Rhod. Agric. Journ. **51**: 501 (1954).—Chippindall in Meredith, Grasses & Pastures of S. Afr.: 316 (1955).—Pilg. in Engl. & Prantl, Nat. Pflanzenfam., ed. 2, **14d**: 123 (1956).—de Winter in Bothalia, **8**: 289, fig. 67, 68, 69, 159/25 (1965). Type as above.

Rhodesia. N: Gokwe, Chimvuri Vlei, 20.ii.1964, *Bingham* 1112 (BM; LISC; SRGH). S: Gwanda, 19·2 km. from Beitbridge, 490 m., 4–5.i.1956, *Rattray* 1751 (SRGH). **Mozambique.** SS: 31·2 km. from the crossing between S. Paulo de Messano and Guijá, 28.iii.1954, *Barbosa & Balsinhas* 5494 (LISC; LMA).
Also in the Transvaal. On sandy soils in open bush savanna.

Unfortunately the holotype of *A. argentea* described by Schweickerdt from the Transvaal was not available for examination during the course of my study. Judging from its description and from some material from the Transvaal, referred to by Schweickerdt to *A. argentea*, it seems that the only difference between *A. argentea* and *A. mollissima* exists in the general appearance of the panicle (as stated in the key above). The characters of the spikelet are very variable in both, and the measurements of its structures overlap. These taxa, therefore, cannot be regarded as distinct spp. As *A. argentea* is mainly confined to eastern areas, as opposed to *A. mollissima* which has a more western distributional range, it seems to be justifiable to treat *A. argentea* as a subsp. of *A. mollissima*. As indicated by Schweickerdt subsp. *argentea* shows a close affinity with *A. stipitata* subsp. *graciliflora*, similar to that existing between *A. mollissima* subsp. *mollissima* and *A. stipitata* subsp. *stipitata*, both pairs of taxa having common characteristics in the panicle and spikelets, the main difference being the presence or absence of dense woolliness on the culms. Both spp., *A. mollissima* and *A. stipitata*, therefore, are still in need of further critical study, involving observations in the field, cultivation—and hybridization—experiments.

25. **Aristida stipitata** Hack. apud Schinz in Verh. Bot. Ver. Prov. Brandenb. **30**: 143 (1888); in Bull. Herb. Boiss. **4**, Append. 3: 19 (1896).—Henrard, Crit. Rev. Aristida **3** in Meded. Rijks-Herb. **54b**: 590 (1928); Monogr. Aristida **1** in Meded. Rijks-Herb. **58**: 106, t. 35 (1929).—Schweickerdt in Bothalia, **4**: 171 (1941).—Chippindall in Meredith, Grasses & Pastures of S. Afr.: 316 (1955).—de Winter in Bothalia, **8**: 290, fig. 70, 159/26 (1965).—Launert in Merxm., Prodr. Fl. SW. Afr. **160**: 34 (1970). Type from SW. Africa.

Perennial, 70–150 cm. high, laxly or densely caespitose. Culms erect, robust, usually simple or sparingly branched from the upper nodes, glabrous, smooth, 4–6-noded; nodes glabrous, smooth. Leaf-sheaths usually lax, usually glabrous, smooth, sometimes the lower ones with long scattered hairs. Ligule an obsolete, minutely ciliolate rim; auricles usually short-ciliolate; collar glabrous. Leaf-laminae up to 30 × 0·4 cm., but often much shorter, flat or involute, usually glaucous, minutely scabrous above, glabrous beneath. Panicle 20–25 cm. long, shortly exserted, usually erect and stiff, narrow, dense, spiciform, many-flowered; axis terete, usually glabrous, smooth; branches more or less terete, appressed, glabrous, smooth, or slightly scaberulous. Spikelets pallid or greenish. Glumes unequal, chartaceous, 1-nerved, glabrous, smooth, slightly scaberulous on the keels; the inferior up to 13 mm. long, lanceolate, acuminate, shortly awned; the superior up to 22 mm. long, narrower, gradually tapering into an awn up to 5 mm. long or with a very short awn from the sinus between 2 hyaline lacerate lobes. Lemma 8–9 mm. long, linear or tubular, pallid or dark-mottled, finely punctulate; callus 2–3 mm. long, acuminate, densely barbate; column 20–40 mm. long, strongly twisted, scabrous; awns 25–60 mm. long, subequal, very fine, scabrous; articulation between the apex of the lemma and the base of the column.

This very polymorphic species is so extremely variable that it is rather difficult to distinguish well-marked subspecies. Although its extreme variants are so distinct that several have previously been described as species, yet in many cases they are connected by intergrading forms. When more of these forms can be examined cytogenetically, observed under uniform conditions of cultivation and used in hybridization experiments, a more satisfactory classification will be possible. For the present it is proposed to recognize groups as subspecies on the basis of a general tendency in each group to exhibit a certain combination of characters, accompanied by peculiarities in their distribution.

Panicles narrow, dense, spiciform, with appressed branches and with mostly pallid
   spikelets; auricles of the leaves usually short-ciliolate:
   Plants robust, 70–150 cm. high; culms simple or rarely branched from the upper
     nodes; panicle terminal; column of the awns more than 20 mm. (up to 35 mm.)
     long - - - - - - - - - - - - subsp. *stipitata*
   Plants slender, usually not exceeding 60 cm. in height; culms much branched from the
     upper nodes; panicles borne at the ends on the branches; column of the awns
     15–20 mm. long - - - - - - - - - subsp. *spicata*
Panicles narrow but lax and interrupted, with suberect or spreading branches; spikelets
   often purplish; auricles of the leaves short-ciliate or barbate:
   Plants robust, 70–120 cm. high; culms usually much branched from the upper nodes;
     panicle 20–30 cm. long; column of the awns 10–20 cm. long:
     Leaf-laminae up to 5 mm. wide at the base; auricles of the leaves barbate (with
      usually reflexed or spreading hairs); branches of the panicle spreading and rather
      lax at maturity; column of the awns 15–20 mm. long - - subsp. *robusta*
     Leaf-laminae less than 5 mm. wide at the base; auricles of the leaves short-ciliolate;
      branches of the panicle suberect at maturity; column of the awns 10–12 mm.
      long - - - - - - - - - - subsp. *ramifera*
   Plants slender, 30–60 cm. high; culms sparingly branched from the upper nodes or
     simple; panicle usually 10–20 cm. long, laxly branched, with more or less spreading
     branches; column of the awns 15–20 cm. long - - - subsp. *graciliflora*

Subsp. **stipitata**

Botswana. SW: 107·2 km. N. of Kang on road to Ghanzi, 19.ii.1960, *de Winter* 7393 (K; SRGH). **Zambia.** B: Kalabo–Sikongo road, 15.xi.1959, *Drummond & Cookson* 6503 (BM; SRGH). S: Livingstone Distr., Dambwa Forest Reserve, 15.iv.1963, *Bainbridge* 769 (BM). **Rhodesia.** C: Salisbury Distr., McIlwaine National Park, 1350 m., 13.iii.1965, *Crook* 712 (BM; PRE; SRGH). S: Fort Victoria, Makoholi Experimental Station, 15.i.1948, *Newton & Juliasi* 18 (immature) (SRGH).

With hairy lower leaf-sheaths:

Zambia. S: Machili, 17.ii.1961, *Fanshawe* 6277 (SRGH).
Also in SW. Africa, Orange Free State and the Transvaal. Usually on deep white sandy soils in open savanna-woodland, grass-plain, on dunes covered with grasses and edges of grassy vleis.

According to Henrard (1928), this subsp. is characterized by the column of the awns being very long, always more than 35 mm. The column in the specimens occurring in the Flora Zambesiaca area, however, is usually shorter than 35 mm. but exceeding 20 mm. in length. This subsp. has a mainly south-western distribution in S. Africa, being rather rare in its eastern parts. When the areas of subsp. *stipitata* and subsp. *robusta* overlap or meet, forms having some characters intermediate between these subsp. are not uncommon. They have characteristics of the panicle of subsp. *stipitata* but the auricles of their leaves are barbate and the column of the awns does not exceed 20 mm., as in subsp. *robusta*. Such transitional forms are met with in the following areas:—

Botswana. N: 77 km. N. of Aha Hills on SW. African border, 13.iii.1965, *Wild & Drummond* 6989 (BM; SRGH). **Zambia.** B: Sesheke, i.1924, *Borle* 11 (K, SRGH). **Rhodesia.** W: Nyamandhlovu, 64 km. from Bulawayo on Falls road, 5–9.iii.1956, *Rattray* 1764 (SRGH).

Subsp. **robusta** (Stent & Rattray) Melderis apud Launert in Merxm., Prodr. Fl. SW. Afr. **160**: 34 (1970).
   Syntypes: Rhodesia, Wankie, near Fuller Siding, Katsetchetti Vlei, *Pardy* in GHS 3875; Nyamandhlovu, Gwaai Forest Reserve, *Pardy* in GHS 4071 (SRGH, syntypes).
   Aristida graciliflora var. *robusta* Stent & Rattray in Proc. Rhod. Sci. Ass. **32**: 44 (1933).—Sturgeon in Rhod. Agric. Journ. **51**: 501 (1954). Syntypes as above.
   Aristida wachteri Henrard, Crit. Rev. Aristida, Suppl. in Meded. Rijks-Herb. **54c**: 746 (1933).—Sturgeon in Rhod. Agric. Journ. **51**: 501 (1954). Type: Rhodesia, Nyamandhlovu Distr., Umgusa Spur, Matundhlamahla Vlei, *Pardy* 53 (GHS 3744) (L, holotype; K; SRGH, isotypes).
   Aristida stipitata var. *robusta* (Stent & Rattray) de Winter in Kirkia, **3**: 133 (1963); in Bothalia, **8**: 291, fig. 70 (1965). Syntypes as for *Aristida stipitata* subsp. *robusta*.

Botswana. N: 3 km. S. of Maun-Bushman Pits road on Makalamabedi road, 21.iii.1965, *Wild & Drummond* 7194 (BM; SRGH). SW: 107·2 km. N. of Kang on road to Ghanzi, 19.ii.1960, *de Winter* 7394 (SRGH). SE: Kweneng, between Botlapatlou and Ngware, 3.vi.1956, *de Beer* K5 (LISC; SRGH). **Zambia.** B: Bulazi, 7.i.1960, *Gilges* 868 (SRGH). S: Livingstone, 24.iv.1960, *Fanshawe* 5641 (SRGH). **Rhodesia.** N: Gokwe, 16.v.1962, *Bingham* 258 (BM; SRGH). W: Shangani Distr., Gwampa Forest Reserve, iii.1955, *Goldsmith* 139 (K; LISC; PRE; SRGH). C: Salisbury Distr., Hunyani, S. bank, 24.xi.1929, *Eyles* 6897 (SRGH; K).
Also in SW. Africa. On deep white and red Kalahari sand, in *Terminalia sericea* scrub, woodland, sand-dunes covered with bush and grass, heavy sandveld, amongst rocks on old quartzite hills and on disturbed ground.

A comparison of the isotypes of *A. wachteri* from SRGH and K with syntypes of subsp. *robusta* shows that *A. wachteri* differs from the latter mainly in having hairy axils of the branches in the panicle. As this feature occurs in subsp. *robusta* as well as in other subspecies of this species, and is not correlated with other more constant characters, *A. wachteri* is included here in subsp. *robusta*. In habit this subspecies resembles subsp. *stipitata*, but the branches of its panicle are lax and spreading at maturity, the auricles of the leaves are conspicuously barbate and the column of the awns does not exceed 20 mm. in length. Intermediates between subsp. *robusta* and subsp. *stipitata* are not rare (see subsp. *stipitata*).

Subsp. **spicata** (de Winter) Melderis apud Launert in Merxm., Prodr. Fl. SW. Afr. **160**: 34 (1970). Type: Botswana, Tsabong, *de Winter* 7485 (PRE, holotype; K; SRGH, isotypes).
   Aristida stipitata var. *spicata* de Winter in Kirkia, **3**: 133 (1963); in Bothalia, **8**: 292, fig. 70 (1965). Type as above.

Botswana. N: Ngamiland, 64 km. E. of Maun, 940 m., iii.1958, *Robertson* 619 (LISC; SRGH). SE: Tsabong, near Tsabong Camp, 3.iii.1958, *de Beer* 709 (BM; SRGH).

With barbate auricles of the leaves:
Zambia: Gonge Falls, 1020 m., 27.vii.1952, *Codd* 7210 (PRE).
This subsp. is characterized by having glaucous culms and leaves. The culms rarely exceed 60 cm. in height, being usually much branched, especially in their upper part. The panicles are pallid, narrow, dense, spiciform, terminal and borne at the ends of the branches. The auricles of the leaves are usually short-ciliolate, but sometimes possessing stiff and spreading hairs as in subsp. *robusta*, indicating a close relationship to the latter.

Subsp. **ramifera** (Pilg.) Melderis in Bol. Soc. Brot., Sér. 2, **44**: 288 (1970). Type: Mozambique, Delagoa Bay, in thicket, *Schlechter* 11966 (B, holotype; BM; COI; K, isotypes).
   Aristida ramifera Pilg. in Engl., Bot. Jahrb. **39**: 599 (1907); in R.E. Fr., Wiss. Ergebn. Schwed. Rhod.-Kongo Exped. 1911–1912, **1**: 206 (1916).—Henrard, Crit. Rev. Aristida **3** in Meded. Rijks-Herb. **54b**: 492 (1928); Monogr. Aristida **1** in Meded. Rijks-Herb. **58**: 115, t. 40 (1929). Type as above.

Zambia. S: Livingstone, 900 m., iv.1909, *Rogers* 7048 (K). **Rhodesia.** C: Salisbury, 27.xi.1929, *Eyles* 6897 (K; SRGH). **Mozambique.** LM: Sábiè Pessene region, Moamba, 19.ii.1948, *Torre* 7369 (LISC).
Not known from elsewhere. On sandy ground in thickets and open situations.

This subsp. is characterized by an erect fairly dense inflorescence with suberect branches, bearing spikelets to the base and a rather short column of the awns, 10–12 mm. long, differing from the other subspecies in these characters.

Subsp. **graciliflora** (Pilg.) Melderis apud Launert in Merxm., Prodr. Fl. SW. Afr. **160**: 34 (1970). Type: Mozambique, Delagoa Bay, dunes, *Schlechter* 11984 (B, holotype; BM; COI, isotypes).
  Aristida graciliflora Pilg. in Engl., Bot. Jahrb. **39**: 599 (1907).—Henrard, Crit. Rev. Aristida 1 in Meded. Rijks-Herb. **54**: 211 (1926); Monogr. Aristida 1 in Meded. Rijks-Herb. **58**: 113, t. 40 (1929).—Stent & Rattray in Proc. Rhod. Sci. Ass. **32**: 44 (1933).—Schweickerdt in Bothalia, **4**: 171 (1941).—Sturgeon in Rhod. Agric. Journ. **51**: 501 (1954).—Chippindall in Meredith, Grasses & Pastures of S. Afr.: 317, map 46 (1955). Type as above.
  Aristida vinosa Henrard, Crit. Rev. Aristida Suppl., op. cit. **54c**: 745 (1933).—Sturgeon, tom. cit.: 502 (1954). Type: Rhodesia, Matopos Distr., Matobo, *Rattray* 264 (L, holotype; SRGH, isotype).
  2n = 22.

Botswana. SE: Mahalapye, Morale, 980 m., iii.1957, *de Beer* MP.15 (SRGH). **Rhodesia.** W: Plumtree, 1388 m., 8.iii.1930, *Brain* 555 (BM; K; SRGH). C: Salisbury, 1460 m., 14.iv.1921, *Eyles* 2987 (SRGH). S: Beitbridge, Chiturupazi, 25.ii.1961, *Wild* 5395 (BM; SRGH). **Mozambique.** SS: between Chedenguel and Inharrime, 10.xii.1944, *Mendonça* 3360 (LISC; SRGH). LM: between Marracuene and Alvor, 5.vi.1947, *Barbosa & Lemos* 7596 (COI; LISC; LMA).

With barbate auricles of the leaves:
Rhodesia. W: Gwaai Special Native Area, 1050 m., iv.1953, *Davies* 478 (SRGH). S: Gwanda, c. 24 km. from Beitbridge, on paragneiss, 480 m., 4–5.i.1956, *Rattray* 1737 (SRGH). **Mozambique.** LM: Magude, neighbourhood of Mapulanguene, 30.xi.1944, *Mendonça* 3185 (LISC).

With pubescent leaf-sheaths and barbate auricles of the leaves:
Botswana. SE: Mahalapye, Morale, 980 m., xii.1955, *de Beer* 13 (K; SRGH). **Rhodesia.** W: Bulalima Mangwe Distr., Nata Reserve, SE. area, 12.iv.1951, *Robinson* 387 (SRGH). S: Gutu, Chikwanda Reserve near Mtilikwe R., 24.i.1950, *Cleghorn* 20 (K; SRGH). **Mozambique.** LM: neighbourhood of Lourenço Marques, xii.1945, *Pimenta* ?52530 (LISC).

With unbranched culms:
Botswana. SE: Gaberones, 22·4 km. N. of Lobatsi on Great North Road, 15.xi.1948, *Robertson* 565 (SRGH). **Zambia.** S: Mazabuka, Siamambo stream 3·2 km. below confluence with Bunchele, 3.iii.1960, *White* 7569 (SRGH). **Rhodesia.** W: Matobo, Matopos Research Station, ii.1960, *Rattray* 1826 (BM; SRGH) S: Makoholi Experimental Station, Fort Victoria, ii–iii.1945, *Clarke* 11 (SRGH).
Also in the Cape Prov., Orange Free State, the Transvaal, Swaziland and Natal. Often on sandy and red loamy soils, also on paragneiss, in sandveld, savanna, dry *Mopane* bushland, in seepage zones, on disturbed ground and on dunes and in rocky situations.

Examination of the isotype of *A. vinosa* has revealed that it differs from *A. stipitata* subsp. *graciliflora* only in having unbranched culms. Since the branching of the culms cannot be considered a good diagnostic character, *A. vinosa* is regarded in this treatment as synonymous with subsp. *graciliflora*. The latter can be readily distinguished from the other subspecies of *A. stipitata* in having narrow, but lax, usually effuse panicles with more or less spreading branches bearing usually purplish spikelets. In the characteristics of the leaf anatomy, panicle and spikelets *A. stipitata* shows a marked similarity to *A. mollissima* differing mainly in the absence of pubescence on the culms.

26. **Aristida pilgeri** Henrard, Crit. Rev. Aristida, 2 in Meded. Rijks-Herb. **54a**: 447 (1927); Monogr. Aristida 1 in Meded. Rijks-Herb. **58**: 123, t. 45 (1929).—Schweickerdt in Bothalia, **4**: 156 (1941).—Sturgeon in Rhod. Agric. Journ. **51**: 502 (1954).—Chippindall in Meredith, Grasses & Pastures of S. Afr.: 314 (1955).—de Winter in Bothalia, **8**: 293, fig. 71, 72, 73, 159/27 (1965).—Launert in Merxm., Prodr. Fl. SW. Afr. **160**: 33 (1970). Type from SW. Africa.

Perennial, up to 1·5 m. high, fairly robust, caespitose. Culms erect, glaucous, glabrous, 3–6-noded; nodes glabrous. Leaf-sheaths tight, ciliolate along the margins. Ligule obsolete, minutely ciliate; auricles and the collar glabrous or barbate. Leaf-laminae up to 35 × 0·1–0·2 cm., linear, convolute towards the apex, scabrous above, glabrous and prominently ribbed beneath. Panicle up to 40 × 5 cm., dense, more or less contracted, much branched, with a terete or subangular axis; branches 5–11 cm. long, solitary, erect or nearly appressed, branchlets appressed, scabrous. Spikelets congested, very shortly pedicelled, pallid. Glumes slightly unequal, lanceolate to linear-lanceolate, awned; the inferior 6–7 mm. long, 3-nerved, scaberulous in the upper part, scabrous on the nerves and on the keel, abruptly narrowed into a scabrous awn c. 1·5–2 mm. long; the superior c. 10 mm. long, with a smooth keel, distinctly 2-dentate at the apex with an awn, c. 2 mm. long, from the sinus. Lemma 6–7 mm. long, finely punctulate, scaberulous towards the apex; callus c. 1–1·5 mm. long, obtuse or subobtuse, long-barbate; column very short, c. 1·5 mm. long, slightly twisted; awns subequal, 10–20 mm. long; articulation between apex of the column of the lemma and the base of the awns.

Caprivi Strip. Lisikili, 24 km. E. of Katima Mulilo, 980 m., 17.vii.1952, *Codd* 7097 (K; SRGH). **Botswana.** N: Nata R., Makarikari perimeter, 208 km. on Francistown–Maun road, 910 m., 9.iii.1961, *Vesey-FitzGerald* 3171 (BM; SRGH). **Zambia.** B: 3·2 km. W. of Mongu, on Lizulu, 20.v.1964, *Verboom* 1196 (K). S: Kalomo, Bilili Hot Springs, 3.vi.1961, *Mitchell* 7/59 (K; SRGH). **Rhodesia.** N: Kariba, Sebungwe–Kariba road, vi.1956, *Davies* 1953 (SRGH). W: Shangani Distr., Jibi Jibi Dam, 1070 m., iii.1954, *Davies* 773 (SRGH). C: Que Que, ? Gattari Ranch, 19.iii.1964, *West* 4781 (BM; SRGH). S: Buhera Distr., Zone D3, 22.viii.1959, *Cleghorn* 554 (SRGH). Also in SW. Africa and the Transvaal. On water-logged granite-sand, on Kalahari sand, red sandy loams in sandveld grassland and open woodland, grassy depressions at edges of lakes and on sand dunes.

This is a fairly uniform readily distinguishable sp. belonging to a group of the species which has an articulation between the apex of the column and the base of the awns. It is more robust than the other spp. of this group, often reaching 150 cm. in height. Its inflorescence is contracted and much branched, with solitary, up to 11 cm. long, more or less appressed branches and appressed spikelets. The column of the lemma is shorter than in the other spp., being c. 1·5 mm. long. In *A. congesta*, the panicle is spiciform, very dense or narrow and interrupted, with very short branches, but in *A. barbicollis*, it has spreading branches, bearing clusters of spikelets towards the ends of the branches.

27. **Aristida congesta** Roem. & Schult. in L. Syst. Veg., ed. nov., **2**: 401 (1817).—Kunth, Enum. Pl. **1**: 195 (1833).—Trin. & Rupr., Sp. Gram. Stip.: 153 (1842).—Steud., Syn. Pl. Glum. **1**: 142 (1854).—Dur. & Schinz, Consp. Fl. Afr. **5**: 802 (1894).—Hack. in Bull. Herb. Boiss. **4**, Append. **3**: 18 (1896).—Stapf in Harv. & Sond., F.C. **7**: 558 (1899).—Monro in Proc. Rhod. Sci. Assoc. **6**: 37 (1906).—Henrard, Crit. Rev. Aristida **1** in Meded. Rijks-Herb. **54**: 113 (1926); Monogr. Aristida **1** in Meded. Rijks-Herb. **58**: 126, t. 47 (1929).—Stent & Rattray in Proc. Rhod. Sci. Ass. **32**: 45 (1933).—Schweickerdt in Bothalia, **4**: 157 (1941).—Sturgeon in Rhod. Agric. Journ. **51**: 502 (1954).—Chippindall in Meredith, Grasses & Pastures of S. Afr.: 314 fig. 280, map 44 (1955).—de Winter in Bothalia, **8**: 296, fig. 74, 75, 76, 159/28 (1965).—W. D. Clayton, F.T.E.A., Gramineae: 156 (1970).—Launert in Merxm., Prodr. Fl. SW. Afr. **160**: 30 (1970). Type from S. Africa (Cape).
Chaetaria congesta (Roem. & Schult.) Nees, Fl. Afr. Austr.: 189 (1841). Type as above.
Aristida longicauda Hack. in Bol. Soc. Brot. **6**: 143 (1888).—Dur. & Schinz, Consp. Fl. Afr. **5**: 804 (1894).—Henrard, Crit. Rev. Aristida **2** in Meded. Rijks-Herb. **54a**: 305 (1927); Monogr. Aristida **1** in Meded. Rijks-Herb. **58**: 122, t. 44 (1929).—Schweickerdt in Bothalia, **4**: 156 (1941).—Sturgeon in Rhod. Agric. Journ. **51**: 502 (1954).—Chippindall in Meredith, Grasses & Pastures of S. Afr.: 314 (1955). Type: Mozambique, *Carvalho* 35 (W, holotype).
Aristida alopecuroides Hack. apud Schinz in Verh. Bot. Ver., Prov. Brandenb. **30**: 144 (1888); in Bull. Herb. Boiss. **4**, Append. **3**: 17 (1896).—Dur. & Schinz, Consp. Fl. Afr. **5**: 800 (1894).—Henrard Crit. Rev. Aristida **1** in Meded. Rijks-Herb. **54**: 22 (1926); Monogr. Aristida **1** in Meded. Rijks-Herb. **58**: 121, t. 43 (1929).—Schweickerdt in Bothalia, **4**: 155 (1941).—Chippindall in Meredith Grasses & Pastures of S. Afr.: 314 (1955). Type from SW. Africa.
Aristida congesta var. *genuina* Chiov., Fl. Erit.: 333 (1908).—Henrard, op. cit. **54**: 115 (1926). Type from S. Africa (Cape).
Aristida congesta var. *pilifera* Chiov., loc. cit.—Henrard, tom. cit.: 115, (1926). Type from Eritrea.

Aristida rangei Pilg. in Engl., Bot. Jahrb. **48**: 344 (1912).—Henrard, Crit. Rev.
Aristida **3** in Meded. Rijks-Herb. **54b**: 498 (1928); Monogr. Aristida **1** in Meded.
Rijks-Herb. **58**: 121, t. 44 (1929). Type from SW. Africa.
Aristida elytrophoroides Chiov. in L'Agric. Colon. Firenze, **18**: 351 (1924).—
Henrard, Crit. Rev. Aristida **1** in Meded. Rijks-Herb. **54**: 168 (1926); Monogr.
Aristida **1** in Meded. Rijks-Herb. **58**: 124, t. 45 (1929).—Stent & Rattray in Proc.
Rhod. Sci. Ass. **32**: 45 (1933).—Sturgeon in Rhod. Agric. Journ. **51**: 502 (1954).
Type from Eritrea.
Aristida congesta var. *megalostachya* Henrard, Monogr. Aristida **1** in Meded.
Rijks-Herb. **58**: 126 (1929); Crit. Rev. Aristida Suppl. in Meded. Rijks-Herb. **54c**:
711 (1933).—Stent & Rattray, loc. cit.—Sturgeon, op. cit.: 503 (1954). Syntypes:
Rhodesia, Mrewa, i.1911, *Appleton* 26 (K); Bulawayo and Matopo Hills, *Appleton*
17 (K).

Perennial, up to 75(90) cm. high, densely caespitose. Culms erect or geniculate,
simple or branched from the lower or upper nodes, rather wiry, nearly terete,
glabrous, smooth, 3–4-noded; nodes glabrous, smooth. Leaf-sheaths slightly
compressed, striate, glabrous or scaberulous upwards, the lower ones strongly
keeled, with margins more or less overlapping at the base, the upper sometimes
slightly keeled, tight. Ligule a short-ciliolate rim; auricles barbate or glabrous;
collar glabrous or minutely pubescent or somewhat barbate. Leaf-laminae up to
20 cm. long, linear, conduplicate to convolute, usually flat at the base, acuminate,
fairly rigid and more or less curved, scabrous or hispidulous above, glabrous and
smooth beneath, with the margins not thickened. Panicle up to 20(25) cm. long,
dense, contracted, spiciform, often interrupted towards the base, sometimes with 1
or 2 subspicate more or less spreading branches at the base; axis subterete,
scaberulous; branches short, solitary, much divided from the base, usually
appressed, slightly compressed, scaberulous. Spikelets densely fascicled, purplish
or greenish. Glumes unequal, subhyaline; the inferior 4–8 mm. long, lanceolate,
scabrous or scaberulous on the keel and often minutely scaberulous on margins
towards the apex, often abruptly contracted at the apex, with a short awn, 1–4 mm.
long; the superior 5–9 mm. long, narrower, scaberulous on the keel towards the
apex, with a short awn 1–3 mm. long. Lemma 4–6 mm. long, tubular, scabrid,
finely tuberculed, coarsely scabrid towards the apex; callus 1–1·5 mm. long, acute
or acuminate, shortly barbate; column up to 4–6 mm. long, twisted, scabrous or
scaberulous; awns subequal, up to 12–25(30) mm. long, scabrous, more or less
divergent; articulation between the apex of the column and the base of the awns.
2n = 22.

Caprivi Strip. Linyanti Area, c. 80·5 km. from Katima on road to Linyanti, c. 910 m.,
27. xii.1958, *Killick & Leistner* 3138 (PRE; SRGH). **Botswana.** N: Ngamiland, Tsau,
960 m., 19.iii.1961, *Vesey-FitzGerald* 3264 (BM; K; SRGH). SW: 64 km. N. of Kan on
road to Ghanzi, 17.ii.1960, *de Winter* 7364 (K; SRGH). SE: 240 km. NW. of Molepolole,
16.vi.1955, *Story* 4905 (K; SRGH). **Zambia.** S: Kafue, 26.iii.1963, *van Rensburg*
1801 (K). **Rhodesia.** N: Gokwe, Gokwe Golf Course, 11.iii.1963, *Bingham* 494 (BM;
K; SRGH). W: Matobo, farm drainage, 1440 m., xii.1957, *Miller* 4845 (SRGH). C:
Marandellas, Grasslands, 4.vii.1947, *Newton* 82 (K; SRGH). E: Umtali, Carolina
B Farm near Pounsley, 1020 m., 27.ii.1957, *Phipps* 566 (COI; LISC; SRGH). S: Bubi,
on Kalahari sand (near Bunbizwana) R., 24.iii.1931, *Pardy* in GHS 4870 (SRGH).
Mozambique. LM: near Magude, c. 13 km. N. of Mapulanguene, 17.ii.1953, *Myre &*
Balsinhas 1543 (LISC). SS: Régulo ?Bambarine, Manjacaze, 19.iii.1948, *Torre* 7533
(LISC).

Distributed also in NE. Africa, E. Africa southwards to the Cape Prov., Orange Free
State, Lesotho and SW. Africa. On sandy, basalt and black clayey soils, granite outcrops, Kalahari sand overlying karoo basalt and sandstones, in open grassland, vlei,
mopane woodland and paddocks, near hot springs, in fallow fields, on old tobacco lands
and in waste places.

This is a very variable sp. and some variants have been described as separate spp. The
examination of these variants shows that there exists a complete chain of intermediates,
so that it is clearly impossible to draw distinct lines between them. More robust specimens
have been described by Henrard as var. *megalostachya* which grades into *A. congesta*.
A. congesta is closely allied to *A. barbicollis* having a similar structure of the spikelets and
leaf anatomy. In drier areas both have a tendency to become annual. They differ mainly
in the characteristics of the panicle which in *A. congesta* is congested, spiciform, sometimes
with 1 or 2 subspicate branches at the base, while in *A. barbicollis* it is open with spikelets
arranged in spiciform clusters at the ends of the branches.

28. **Aristida barbicollis** Trin. & Rupr., Sp. Gram. Stip.: 152 (1842).—Steud., Syn. Pl. Glum. **1**: 141 (1854).—Dur. & Schinz, Consp. Fl. Afr. **5**: 800 (1894).—Engl., Pflanzenw. Ost-Afr. **C**: 107 (1895).—Stapf. in Harv. & Sond., F.C. **7**: 559 (1899).— Wood, Natal Plants, **5**: pl. 401 (1904).—Rendle in Journ. Linn. Soc. Bot. **40**: 233 (1911).—Henrard, Crit. Rev. Aristida **1** in Meded. Rijks-Herb. **54**: 48 (1926); Monogr. Aristida **1** in Meded. Rijks-Herb. **58**: 132, t. 50 (1929).—Stent & Rattray in Proc. Rhod. Sci. Ass. **32**: 45 (1933).—Bremerk. & Oberm. in Ann. Transv. Mus. **16**: 405 (1935).—Schweickerdt in Bothalia, **4**: 153 (1941).—Sturgeon in Rhod. Agric. Journ. **51**: 503 (1954).—Chippindall in Meredith, Grasses & Pastures of S. Afr.: 313, pl. 9, map 44 (1955).—Pilg. in Engl. & Prantl, Nat. Pflanzenfam. ed. 2, **14d**: 122 (1956).—W. D. Clayton, F.T.E.A., Gramineae: 157, fig. 50 (1970). TAB. **33**. Type from S. Africa (Cape).

Chaetaria forskolii Nees, Fl. Afr. Austr.: 188 (1841) non *Aristida forskohlii* Tausch (1836). Type from S. Africa.

Aristida lommelii Mez in Fedde, Repert. **17**: 150 (1921).—Henrard, Crit. Rev. Aristida **2** in Meded. Rijks-Herb. **54a**: 302 (1927); Monogr. Aristida **1** in Meded. Rijks-Herb. **58**: 133, t. 50 (1929).—Sturgeon in Rhod. Agric. Journ. **51**: 496 (1954).—Robyns & Tournay, Fl. Parc Nat. Alb. **3**: 135, pl. 20 (1955). Type from SW. Africa.

Aristida barbicollis var. *conglomerata* Henrard, Crit. Rev. Aristida Suppl. in Meded. Rijks-Herb. **54c**: 705 (1933). Type from S. Africa.

Aristida congesta subsp. *barbicollis* (Trin. & Rupr.) de Winter in Bothalia, **8**: 173 (1964); Bothalia, **8**: 298, fig. 76, 78, 159/28a (1965). Type as for *A. barbicollis*.

Perennial, 20–65 cm. high, usually glaucous, densely caespitose. Culms erect or ascending, slender, simple or branched from the lower nodes, somewhat wiry, distinctly compressed, especially below, usually glabrous, sometimes slightly scaberulous, 3–4-noded; nodes glabrous, smooth. Leaf-sheaths more or less keeled, tight or lax, glabrous or somewhat scaberulous. Ligule densely short-ciliolate; auricles usually long-barbate; collar laterally barbate, more or less glabrous on the back. Leaf-laminae up to 10(20) cm. long, narrowly linear, conduplicate or convolute, fairly rigid, curved or flexuous, with a rather obtuse apex, with somewhat thickened marginal nerves, scabrous or hispidulous above, glabrous and smooth beneath. Panicle variable in shape and size, usually 15 × 5–10 cm., open, ovate to oblong, with several more or less spreading branches having spikelets congested towards the ends, forming often false spikes; axis erect or flexuous, terete, glabrous, smooth below, more or less subangular, scaberulous in the upper part; branches solitary or 2-nate, distant, scaberulous, naked for 1–6 cm. in the lower part. Spikelets greenish, usually tinged with purple. Glumes keeled, subhyaline, 1-nerved; the inferior usually 5–5·5 mm. long, lanceolate, scaberulous on the keel and laterally towards the apex, usually gradually tapering into a short awn c. 0·8–2 mm. long; the superior c. 6·5–8 mm. long, narrowly lanceolate, usually smooth on the keel, awned or deeply 2-fid at the apex with a c. 1 mm. long awn from the sinus and sometimes with well-developed narrow lateral lobes. Lemma 3·5–4·5 mm. long, sometimes mottled with purple, glabrous below, scabrous towards the apex; callus c. 1 mm. long, usually subacute, densely barbate; column of the awns c. 2·5 mm. long, twisted, scaberulous; awns 11–25 mm. long, subequal or the central one somewhat longer up to 30 mm., slender, scaberulous, more or less spreading; articulation between the apex of the column and the base of the awns.

Botswana. SE: Mahalapye, 15.iv.1931, *Pole Evans* 3206 (SRGH). **Zambia.** S: Mazabuka Distr., Zambezi escarpment 40 km. from Chirundu, layby on Lusaka–Chirundu road, c. 750 m., 3.i.1968, *Simon* 1631 (BM; SRGH). **Rhodesia.** W: Matopo Research Station, 1350 m., 16.ii.1954, *Rattray* 1617 (SRGH). C: Salisbury, Gwebi, 21.xii.1951, *Barnes* 22 (SRGH). E: Umtali Commonage, 1050 m., 3.i.1947, *Fisher* 1177 (SRGH). S: Fort Victoria, Makoholi Experimental Station, *Newton & Juliasi* 13 (SRGH). **Mozambique.** N: Nampula, vii. 1955, without collector 12? (LISC). Z: Between Pebane and Mocubela, 25.x.1942, *Torre* 4684 (LISC). MS: Manica, Macequece, right bank of R. Revuè, 12.iii.1948, *Garcia* 583 (LISC). SS: Chibuto, road to Alto Changane, at 15 km. from Chibuto, Lagoa Culela, 10.vi.1960, *Lemos & Balsinhas* 63 (COI; LISC; LMA).

Also in SW. Africa, Cape Prov., Natal, the Transvaal, Swaziland, Congo, Tanzania and Kenya. On red sandy soils and sandy-clay, paragneiss and serpentine soils and on low granite kopjies projecting through black soils, in sandveld, open bushland, *Acacia* savanna, *Brachystegia* and *Mopane* woodland, on grassy hillsides, along roadsides, in heavily overgrazed old fallows, paddocks, on edges of sugar-cane lands and on disturbed areas.

E

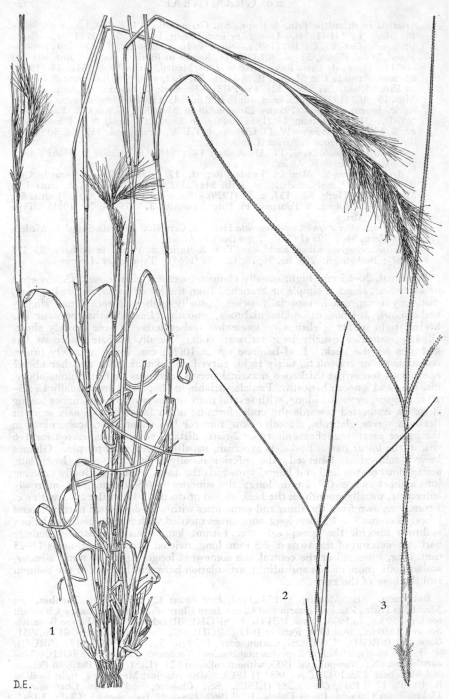

Tab. 33. ARISTIDA BARBICOLLIS. 1, habit (×½); 2, spikelet (×4½); 3, lemma in lateral view (×9), all from *Leippert* 5899.

A very variable sp., closely related to *A. congesta* from which it differs mainly in having a lax and open panicle with ± spreading branches bearing spikelets congested in dense or lax fascicles towards the ends of the branches. As already mentioned by Schweickerdt, intermediates between these spp. are frequent, especially in areas where they both occur. They may have arisen by hybridization. A variant of *A. barbicollis* with longer awns and column on the lemma and a shorter upper glume was described by Mez as a separate sp., *A. lommelii*. According to Henrard, in *A. barbicollis* the lemma, including the callus and column, is 6·5 mm. long, the awns of the lemma 11–12 mm. long, the inferior glume with an awn up to 0·75 mm. long and the superior glume 8 mm. long, whilst in *A. lommelii* the lemma, including the callus and column, is 22–27(30) mm. long, the lower glume with an awn up to 2 mm. long and the upper glume 6·5 mm. long. Examination of a large number of specimens from our area has revealed that the length of awns and column, as well as the size of the glumes are very variable characters, not correlated with other features. Since transitional forms exist between these taxa, their distinction is difficult, and therefore, they are treated here as being conspecific.

29. STIPAGROSTIS Nees

Stipagrostis Nees in Linnaea, **7**: 290 (1832); Fl. Afr. Austr.: 171 (1841).— de Winter in Kirkia, **3**: 133 (1963); in Bothalia, **8**: 307 (1965).

Schistachne Fig. & De Not. in Mem. R. Acad. Sci. Tor. Ser. 2, **12**: 252 (1851).

Spikelets 1-flowered, pedicelled, in terminal contracted or lax panicles. Rhachilla disarticulating above the glumes. Florets hermaphrodite, equalling or shorter than the glumes. Glumes persistent, equal or unequal, narrow, acuminate to obtuse, occasionally emarginate, 1–11-nerved, usually both 3-nerved. Lemma narrow, cylindric, with involute margins, 3-nerved, grooved ventrally, becoming indurated at maturity, bearing at the base a well developed pungent or minutely 2-fid, usually oblique, barbate or glabrous callus, awned from the apex of the lemma or from the apex of a usually twisted column; awns usually 3, all plumose, or at least the central one plumose, or with a penicil of hairs at the apex of the column; articulation present, being situated at or above the middle or between the base of the column and the body of the lemma. Palea usually much less than ½ the length of the lemma, indurated, 2-nerved, glabrous. Lodicules usually 2, or absent, obtuse. Stamens 3. Ovary glabrous; styles free; stigmas 2, plumose. Caryopsis terete, frequently shallowly grooved ventrally, tightly enclosed by the lemma, free; hilum linear; embryo usually ⅓–½ the length of the caryopsis. Annual or densely caespitose perennial grasses with a well developed knotty rhizomatous base, with usually erect, simple or branched, hollow or solid culms, and with usually long narrow leaves; ligule of a dense rim of cilia; panicles narrow, spiciform or effuse and open.

A genus of c. 50 spp., occurring in the deserts or semideserts of the Old World, in Africa, the Volga region of the European U.S.S.R., in the Middle East to Western Tibet, in Central Asia and in Western Siberia.

In Henrard's " Monograph of the Genus Aristida " (1929–1932) the genus *Stipagrostis* was included under separate sections: *Schistachne* Henrard and *Stipagrostis* (Nees) Trin. & Rupr. De Winter in his works on South African *Aristideae* (1963, 1965) has followed Nees by keeping *Stipagrostis* as a separate genus apart from *Aristida*. As a differential character of the external morphology of *Stipagrostis* he has given the pubescence of the awns of the lemma, which are plumose or at least the central one plumose, or if the latter is glabrous and single, then with a penicil of hairs at the apex of the column. According to de Winter the spp. belonging to *Stipagrostis* differ from those of *Aristida* in having a different leaf-anatomy; the outer bundle sheath, consisting of larger (or at least not smaller) cells than the inner bundle sheaths, has chloroplasts which are absent in the inner one; the cells of the inner bundle sheaths being fairly thick-walled and more lignified than the outer ones; and the silicified cells of the epidermis being usually subcircular in shape.

Nodes of the culms conspicuously barbate, having a ring of spreading white hairs; articulation near the middle of the lemma (at maturity awns and column breaking off with a conical, hollow, upper portion of the lemma attached) -   -   -   1. *ciliata*
Nodes of the culms glabrous; articulation at or slightly below the base of the column (at maturity awns and column breaking off without or with only a very short apical portion of the lemma attached):

Glumes usually glabrous, smooth or finely scaberulous; callus of the lemma densely
  barbate, with hairs not arranged in 2 rings around the callus; central awn naked
  in the lower ½, plumose to the apex; penicil of hairs at the base of the awns absent
  or present:
Inferior glume usually exceeding the superior in length; penicil of hairs at the base
  of the awns absent; leaf-laminae with an obtuse apex   -   -   2. *obtusa*
Inferior glume exceeded by the superior in length; penicil of hairs at the base of the
  awns present; leaf-laminae with an acuminate apex -   -   -   3. *uniplumis*
Glumes pilose all over or at least on the back; callus of the lemma densely barbate, with
  hairs arranged into 2 distant rings around the callus (1 at the base, the other at the
  apex); central awn plumose, either from the upper part of the column, or from the
  base, or only in the upper part, with a naked apex; penicil of hairs at the base of
  the awns absent -   -   -   -   -   -   -   -   -   4. *hirtigluma*

1. **Stipagrostis ciliata** (Desf.) de Winter in Kirkia, **3**: 133 (1963).—Bor, Fl. of Iraq,
   9: 386, pl. 147 (1968); Fl. Iran. **70**: 375, t. 55 (1970).—Launert in Merxm., Prodr.
   Fl. SW. Afr. **160**: 195 (1970). Type from Tunisia.
       Aristida ciliata Desf. in Schrad., Neues Journ. Bot. **3**: 255 (1809).—Del., Fl.
   Égypte Expl. Planches, **2**: 175, pl. 13, fig. 3 (1813).—Trin. & Rupr., Sp. Gram.
   Stip.: 163 (1842).—Steud., Syn. Pl. Glum. **1**: 143 (1854).—Boiss., Fl. Orient. **5**:
   494 (1884).—Dur. & Schinz, Consp. Fl. Afr. **5**: 801 (1894).—Henrard, Crit. Rev.
   Aristida **1** in Meded. Rijks-Herb. **54**: 91 (1926); Monogr. Aristida **1** in Meded.
   Rijks-Herb. **58**: 42, t. 2 (1929).—Post, Fl. Syr., Palest. & Sinai, **2**: 722 (1933).—
   Blatter, Fl. Arab. **5** in Rec. Bot. Surv. of India, **8**: 496 (1936).—Osk. Schwarz,
   Fl. Trop. Arabia: 319 (1939).—V. & G. Täckholm & Drar, Fl. Egypt, **1**: 361 (1941).
   —Maire, Fl. Afr. Nord. **2**: 32, fig. 223 (1953).—Pilg. in Engl. & Prantl, Nat. Pflan-
   zenfam. ed. 2, **14d**: 134 (1956).—Bor, Grass. of B.C.I. & P.: 409 (1960). Type as
   above.
       Arthratherum ciliatum (Desf.) Nees in Linnaea, **7**: 289 (1832). Type as above.
       Schistachne ciliata (Desf.) Fig. & De Not. in Mem. R. Acad. Sci. Tor. Ser. 2, **12**:
   252 (1851).

Perennial up to 85 cm. high, densely or laxly caespitose. Culms erect or geniculately ascending, simple, mostly branched at the base, glabrous, 2–3-noded;
nodes barbate (with long spreading hairs). Leaf-sheaths crowded at the base,
striate, whitish, glabrous or more or less woolly along the margins, sometimes
densely villose, the upper ones tight. Ligule a short-ciliate rim; auricles short-
ciliate, those of innovations long-barbate; collar glabrous. Leaf-laminae variable
in length, convolute, nearly setaceous to pungent, more or less recurved, scabrous or
minutely pubescent or hirtellous above, glabrous and smooth or sparsely pilose
beneath. Panicle up to 30 cm. long, usually narrowly contracted but often open and
sometimes lax, with a nearly glabrous axis, and with erect, appressed, solitary or
usually 2-nate branches. Spikelets linear-oblong, pallid or straw-coloured, often with
purple spot at the base. Glumes nearly equal, 3-nerved, usually glabrous, obtuse or
subacute, subcoriaceous; the inferior 8·5–11 mm. long, the superior 9–12 mm.
long. Lemma shorter than the glumes, cylindric, smooth, articulated near the
middle; callus 1·5–2·5 mm. long, very acute, long-hairy; column 10–13 mm.
long, straight or slightly twisted, filiform; central awn up to 4·5 cm. long, plumose,
more or less naked in the lower ¼, with a short naked exserted apex; the lateral
awns 1·5–2 cm. long, very fine, mostly erect, scaberulous.

Botswana. SW: Tsabong, 25.ii.1960, *Yalala* 84 (SRGH).
Also in SW. and N. Africa, eastwards to Sinai, Arabia and Afghanistan. Usually in
open grassland on dry sandy soils in hilly areas.

S. ciliata is a very variable sp., easily recognized, however, by the conspicuously
spreading hairs on the nodes of the culms, the erect mostly unbranched culms, and by the
subcoriaceous glumes. In the characters of the central awn of the lemma, the specimen
cited above resembles the typical form of *S. ciliata* in which, according to Henrard, the
central awn has a naked excurrent apex and its plumose part is ± subacute in outline.
In S. African specimens, referred to a variety—*S. ciliata* var. *capensis* (Trin. & Rupr.)
de Winter, the central awn is without a naked excurrent apex and its plumose part is
obtuse in outline.

2. **Stipagrostis obtusa** (Del.) Nees in Linnaea, **7**: 293 (1832).—de Winter in Bothalia,
   8: 355, fig. 128, 129, 160/19 (1965).—Bor, Fl. of Iraq, **9**: 388 (1968); Fl. Iran.: 374
   (1970).—Launert in Merxm., Prodr. Fl. SW. Afr. **160**: 199 (1970). Type from Egypt.
       Aristida obtusa Del., Fl. Égypte. Expl. Planches, **2**: 175, t. 13 fig. 2 (1813).—
   Trin. & Rupr., Sp. Gram. Stip.: 167 (1842).—Steud., Syn. Pl. Glum. **1**: 144 (1854).—

Boiss., Fl. Orient. **5**: 494 (1884).—Stapf in Harv. & Sond., F.C. **7**: 567 (1899), pro parte.—Dinter in Fedde, Repert. **15**: 342 (1918).—Henrard, Crit. Rev. Aristida 2 in Meded. Rijks-Herb. **54a**: 387 (1927); Monogr. Aristida **1** in Meded. Rijks-Herb. **58**: 72, t. 17 (1929).—Post, Fl. Syr., Palest. & Sinai, **2**: 722 (1933).—Blatter, Fl. Arab. **5** in Rec. Bot. Surv. of India, **8**: 496 (1936).—Osk. Schwarz, Fl. Trop. Arab.: 320 (1939).—Schweickerdt in Bothalia, **4**: 125 (1941).—V. & G. Täckholm & Drar, Fl. Egypt, **1**: 362 (1941).—Maire, Fl. Afr. Nord, **2**: 47, fig. 233 (1953).—Chippindall in Meredith, Grasses & Pastures of S. Afr.: 303, fig. 269, map 37 (1955).—Pilg. in Engl. & Prantl, Nat. Pflanzenfam. ed. 2, **14d**: 123 (1956).—Bor, Grass. B.C.I. & P.: 411 (1955). Type as above.

Stipagrostis capensis Nees in Linnaea, **7**: 291 (1832). Type from S. Africa.

Arthratherum obtusum (Del.) Nees, Fl. Afr. Austr.: 179 (1841).—Jaub. & Spach, Ill. Pl. Orient. **4**: 58, t. 338 (1851). Type as for *Stipagrostis obtusa*.

Perennial, up to 60 cm. high, densely caespitose, with numerous innovations. Culms erect, slender, simple, glabrous, smooth, 1-noded; nodes dark-coloured, glabrous, smooth. Leaf-sheaths striate, glabrous, smooth. Ligule short-ciliate rim; auricles densely barbate; collar glabrous. Leaf-laminae variable in length, 1·5–25 cm. long, c. 0·5 mm. wide, glaucous, setaceous, convolute, subterete, slightly striate, ending in a very obtuse apex, minutely villous or hirtellous above, glabrous, smooth or finely scaberulous beneath. Panicle 2·5–20 cm. long, very narrow, erect, contracted, but fairly lax and interrupted in the lower part; axis terete, striate, glabrous, smooth in the lower part, scaberulous in the upper; branches solitary, 2-partite nearly from the base, filiform, appressed or the lower suberect, with scaberulous branchlets; axils glabrous. Spikelets yellowish, often tinged with purple at the base. Glumes subequal, more or less keeled, finely scaberulous, at least towards the apex; the inferior 8·5–12 mm. long, usually longer than the superior, 3-nerved, obtuse; the superior 8–11 mm. long, 1–3-nerved, subacute. Lemma 2–2·5 mm. long, oblong-cylindric, glabrous, smooth in the lower part, slightly scaberulous towards the apex; callus c. 1 mm. long, acute, densely barbate; column 4–10 mm. long, erect or slightly twisted, scaberulous; central awn 2–3 cm. long, scaberulous in the lower part, plumose towards the apex; lateral awns 0·7–1·8 cm. long, delicate, slightly scaberulous.

Botswana. SW: 24·1 km. E. of border of SW. Africa, SW. of Ghanzi, 30.vii.1955, *Story* 5073 (BM; K).

Also in N. Africa, Sinai Peninsula, Iraq, E. Pakistan, SW. Africa, the Cape Prov. and Orange Free State. Open dry sandy habitats (deserts or semi-deserts) and on hard soil on the floor of pans.

This species may usually be recognised by its strong root system consisting of long roots covered with a dense layer of woolly epidermal hairs, by dense tufts of short, rigid, setaceous, glaucous, curved leaves with an obtuse apex, by contracted panicles, interrupted at the base, often with suberect lower 1–2 branches, and by pallid spikelets tinged with purple at the base. It is closely allied to *S. uniplumis* from which it may be distinguished by the absence of the penicil of hairs at the base of the awns of the lemma and by the presence of the inferior glume exceeding the superior in length.

In their external morphology the specimens of *S. obtusa* from Botswana agree with the typical form. Forms with long-hairy leaves and a 2-fid callus of the lemma, described by J. Karl in Mitt. Bot. Staatssamml. München, **3**: 86 (1951) from SW. Africa as *Aristida bifida*, and regarded by de Winter (op. cit.: 357) to be minor variants of *S. obtusa*, have not yet been found in the Flora Zambesiaca area. Although de Winter has stated that the shape of the apex of the callus is not a constant character, and that specimens with an acute or 2-fid callus have been collected in the same populations, the status of " *A. bifida* " is still in need of further investigation, especially taking into account that " *A. bifida* " is the only species of *Stipagrostis* which has a 2-fid callus. It seems, therefore, that there might be some justification for keeping forms with a 2-fid callus apart from those with an acute callus. The examination of material of *Stipagrostis* named as *S. obtusa* from S. Africa, SW. Africa and Arabia at the British Museum (Natural History) revealed that the presence of the 2-fid callus is associated with hairiness of the leaves and a taller growth of the plant. In contrast all forms with an acute callus had glabrous and smooth leaves and in general were smaller. It is possible, that in areas where both species grow side by side, they produce hybrids which obscure their delimitation.

3. **Stipagrostis uniplumis** (Licht.) de Winter in Kirkia, **3**: 136 (1963);—in Bothalia, **8**: 359, fig. 130, 131 & 132 (1965).—W. D. Clayton in F.T.E.A., Gramineae: 138, fig. 46

(1970).—Launert in Merxm., Prodr. Fl. SW. Afr. **160**: 201 (1970). TAB. **34.** Type from S. Africa.

Aristida uniplumis Licht. apud Roem. & Schult. in L. Syst. Veg. ed. nov. **2**: 401 (1817).—Trin. & Rupr., Sp. Gram. Stip.: 172 (1842).—Steud., Syn. Pl. Glum. **1**: 144 (1854).—Dur. & Schinz, Consp. Fl. Afr. **5**: 809 (1894).—Henrard, Crit. Rev. Aristida 1 in Meded. Rijks-Herb. **54b**: 643 (1928); Monogr. Aristida 1 in Meded. Rijks-Herb. **58**: 77, t. 19 (1929).—Schweikerdt in Bothalia, **4**: 128 (1941).—Sturgeon in Rhod. Agric. Journ. **51**: 500 (1954).—Chippindall in Meredith, Grasses & Pastures of S. Afr.: 305, fig. 271, map 39 (1955).—Pilg. in Engl. & Prantl, Nat. Pflanzenfam. ed. 2. **14d**: 123 (1956). Type as above.

Aristida uniplumis var. *pearsonii* Henrard, Crit. Rev. Aristida 3 in Meded. Rijks-Herb. **54b**: 647 (1928); Monogr. Aristida 1 in Meded. Rijks-Herb. **58**: 77 (1929). —Schweickerdt in Bothalia, **4**: 131 (1941).—Stent & Rattray in Rhod. Sci. Ass. **32**: 44 (1953). Type from Angola.

Perennial or subperennial, up to 75 cm. or more high, usually densely caespitose. Culms erect, slender, simple or branched in the upper part, smooth or scaberulous, 3–4-noded, nodes glabrous. Leaf-sheaths glabrous; the lower ones reduced to scales, the upper tight, striate. Ligule a short-ciliate rim; auricles densely long-barbate; collar glabrous. Leaf-laminae up to 15 cm. long, convolute, setaceous, curved or flexuous, scaberulous above and with a few long hairs towards the ligule, glabrous beneath. Panicle usually long-exserted, narrow, contracted or effuse; axis terete in the lower part, angular and scaberulous in the upper; branches capillary, nearly glabrous, spreading or suberect, usually solitary but branched near the base. Spikelets yellowish or tinged with purple. Glumes unequal, papery, glabrous or with a few hairs on the margins; the inferior 8-9 mm. long, nearly 3-nerved; the superior 9–11 mm. long, conspicuously nerved. Lemma 2–3 mm. long, cylindric, slightly 2-lobed at the apex, glabrous, slightly tuberculate towards the apex; callus 1 mm. long, acute, densely barbate; column c. 5 mm. long, with a dense penicil at the apex, otherwise glabrous; the central awn up to 2·5 cm. long, usually naked in the lower ⅓ and plumose towards the apex, or plumose to the base; lateral awns up to 12 mm. long, glabrous, capillary.

Panicle with numerous spikelets; glumes less than 10 mm. long; central awn of the lemma slender, often pale in colour, usually plumose in the upper ⅔ - - var. *uniplumis*
Panicle often with a few spikelets; glumes usually exceeding 10 mm. in length; central awn of the lemma fairly rigid, usually dark in colour, plumose to the base   var. *neesii*

Var. **uniplumis**

Subperennial or perennial, laxly caespitose. Panicle contracted or open, with numerous spikelets. Glumes not exceeding 10 mm. in length. Lemma 2–3·5 mm. long, excluding callus. Central awn of the lemma slender, usually pallid, diverging at a sharp angle from the lemma, usually plumose in the upper ⅔, naked towards the base. $2n = 44$.

Caprivi Strip. Linyati, 910 m., 27.xii.1958, *Killick & Leistner* 3155 (PRE; SRGH). **Botswana.** N: Ngamiland, 64 km. E. of Maun, 930 m., iv.1958, *Robertson* 643 (K; SRGH). SW: 84·8 km. N. of Kan on road to Ghanzi, 19.ii.1960, *de Winter* 7378 (K; PRE; SRGH). **Rhodesia.** W: Nymandhlovu, Pasture Research Station, 23.iv.1953, *Plowes* 1615 (K; SRGH). S: Beitbridge, 9·6 km. ENE. of Tuli Police Camp, 23.iii.1959, *Drummond* 5964 (BM; SRGH).

Also in Angola, SW. Africa, the Cape Prov., Orange Free State, the Transvaal, Tanzania, Uganda, Kenya and from Somali Republic to Senegal. On open sandy soil (on shallow red Kalahari sand on basalt outcrop, sandy flats, granite sandveld, in sand pockets over limestone cobbles) and on shallow gravelly basalt soil.

Forms with hairy glumes

Botswana. N: Nata R., Makarikari perimeter, 208 km. Francistown–Maun road, 910 m., 9.iii.1961, *Vesey-FitzGerald* 3170 (BM; SRGH). **Rhodesia.** S: Gwanda, Tuli Research Sub-Station, 780 m., 16.i.1959, *Howden* 6 (SRGH).
On deep sandy soil in scrubby woodland and dry lake flats.

Var. **neesii** (Trin. & Rupr.) de Winter in Kirkia, **3**: 136 (1963); in Bothalia, **8**: 360, fig. 132 (1965). Type from S. Africa (Cape).

Aristida uniplumis var *neesii* Trin. & Rupr., Sp. Gram. Stip.: 173 (1842).—Dur. & Schinz, Consp. Fl. Afr. **5**: 809 (1894).—Dinter in Fedde, Repert. **15**: 343 (1918).—

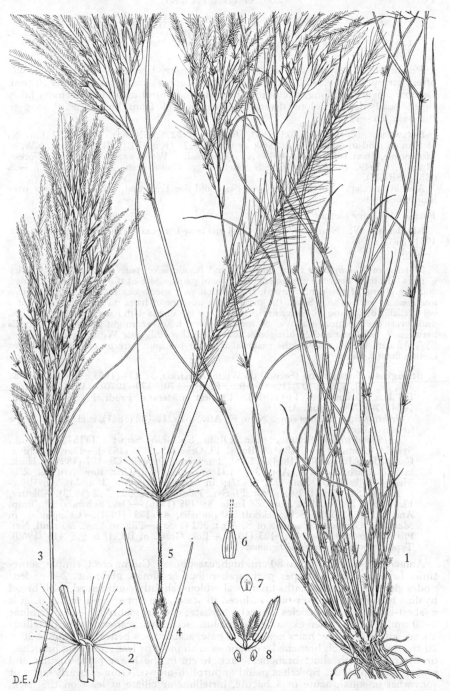

Tab. 34. STIPAGROSTIS UNIPLUMIS. 1, habit (× ⅔); 2, ligule (× 4); 3, inflorescence (× ⅔); 4, glumes (× 4); 5, floret (× 4); 6, lemma (× 4); 7, palea (× 4); 8, sexual organs, with lodicules detached (× 4), all from *Padwa* 217. From F.T.E.A.

Henrard, Crit. Rev. Aristida **3** in Meded. Rijks-Herb. **54b**: 646 (1928); Monogr. Aristida **1** in Meded. Rijks-Herb. **58**: 77 (1929).—Schweickerdt in Bothalia, **4**: 130 (1941).—Chippindall in Meredith, Grasses & Pastures of S. Afr.: 306 (1955). Type as above.

Perennial, densely caespitose, with a short knotty branched rhizome. Panicle often narrow but lax, with a few spikelets. Glumes usually longer than 10 mm. Lemma 3·5–4 mm. long, excluding the callus. Central awn of the lemma fairly rigid, usually dark in colour, diverging at a right angle from the lemma, plumose to the base. 2n = 44.

Botswana. N: Nkate, 20.iv.1931, *Pole Evans* 3298 (SRGH). SW: 53 km. N. of Kan on road to Ghanzi, 19.ii.1960, *de Winter* 7378 (K; PRE). SE: 154 km. NW. of Molepolole, 16.vi.1955, *Story* 4908 (K). **Rhodesia.** W: Gwaai Distr., Gusu Forest, Mpindo, ii.1949, *Davies* 94 (K). S: Beitbridge, Chiturupasi, 25.ii.1961, *Wild* 5408 (BM; SRGH).

Also in the Cape Prov., Orange Free State and the Transvaal. In sand pockets over limestone cobbles.

Forms with hairy glumes

Rhodesia. W: Nymandhlovu, in Gusi vlei near Umgusa, 21.iii.1929, *Pardy* in GHS 4941 (SRGH).

In vleis.

Material of *S. uniplumis* from Botswana and Rhodesia contains also forms with hairy glumes. In their essential characters (the type of panicle, size of the glumes and lemmas, and the plume of the central awn), they match well specimens referred either to var. *uniplumis* or var. *neesii*. Since the presence or absence of hairs on the glumes is not a very reliable diagnostic character in the genus, the specimens with hairy glumes are placed under varieties mentioned above. Some specimens, however, might prove to be hybrids between *S. uniplumis* and *S. hirtigluma* subsp. *gracilior* (Pilg.) de Winter, recorded from SW. Africa. Unfortunately, the material is either too immature or too inadequate for certain determination.

4. **Stipagrostis hirtigluma** (Steud.) de Winter in Kirkia, **3**: 134 (1963); in Bothalia, **8**: 361, fig. 133, 134 & 137 (1965).—Bor, Fl. Iran. **70**: 370 (1970).—W. D. Clayton, F.T.E.A., Gramineae: 140 (1970).—Launert in Merxm., Prodr. Fl. SW. Afr. **160**: 197 (1970). Type from Arabia.

    Arthratherum ciliatum sensu Nees, Fl. Afr. Austr. **1**: 182 (1841), excl. synon. Type from S. Africa.

    Aristida hirtigluma Steud. ex Trin. & Rupr., Sp. Gram. Stip.: 171 (1842).; Steud., Syn. Pl. Glum. **1**: 144 (1854).—Boiss., Fl. Orient. **5**: 496 (1884).—Dur. & Schinz, Consp. Fl. Afr. **5**: 803 (1894).—Engl., Pflanzenw. Ost-Afr. **C**: 107 (1895).—Hack. in Bull. Herb. Boiss. **4**, App. **3**: 18 (1896).—Henrard, Crit. Rev. Aristida, **2** in Meded. Rijks-Herb. **54a**: 231 (1927); in Monogr. Aristida, **1** in Meded. Rijks-Herb. **58**: 68, t. 15 (1929).—Post, Fl. Syr., Palest. & Sinai, **2**: 722 (1933).—Blatter, Fl. Arab. **5** in Rec. Bot. Surv. of India, **8**: 495 (1936).—Osk. Schwarz, Fl. Trop. Arab.: 320 (1939).—Schweickerdt in Bothalia, **4**: 123 (1941).—Chippindall in Meredith, Grasses & Pastures of S. Afr.: 302 (1955).—Pilg. in Engl. & Prantl, Nat. Pflanzenfam. ed. 2., **14d**: 123 (1956).—Bor, Grass. of B.C.I. & P.: 410 (1960). Type as for *Stipagrostis hirtigluma*.

Annual or perennial, up to 80 cm. high, caespitose. Culms erect, simple, some-times branched in the upper part, scaberulous or almost glabrous, 2–5-noded; nodes glabrous. Leaf-sheaths finely scaberulous, slightly compressed, with broad hyaline margins, the lower ones reduced to scales, the upper ones tight. Ligule a short-ciliate rim; auricles densely barbate; collar glabrous. Leaf-laminae 6–20 cm., c. 1 mm. when expanded, convolute, setaceous, scaberulous or hirtellous above, with a few long hairs near to the base, scaberulous beneath. Panicle 30 × 20 cm., lax and much branched, sometimes with more or less spreading branches, or contracted with short branches; axis terete in the lower part, angular and scaberulous upwards. Spikelets pallid or purplish-brown. Glumes nearly equal or somewhat unequal, acute or subacute, hirtellous or ciliate at least on the back, 3-nerved; the inferior 6–11 mm. long, rounded or slightly keeled, abruptly con-tracted or truncate and erose, with a short mucro; the superior 10–13 mm. long, subobtuse or subacute, slightly 2-fid or emarginate, with a short awn from the sinus. Lemma c. 3·5–4 mm. long (including the callus), conspicuously asperulous-tuberculate, especially towards the apex, tapering gradually into a column; callus 0·5–0·75 mm. long, curved, with an oblique, glabrous, long and acute point,

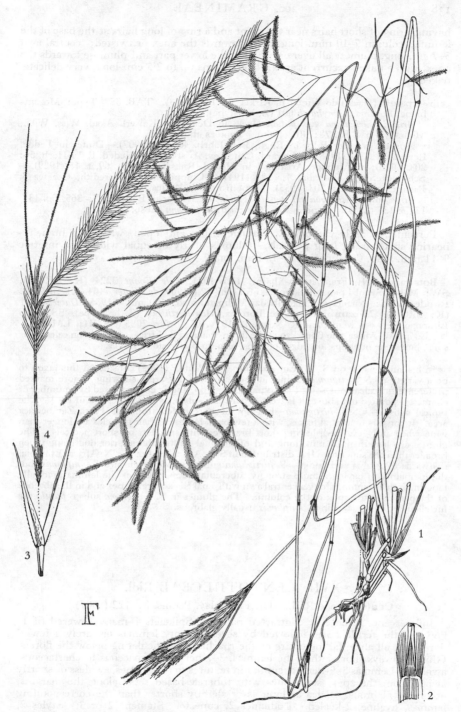

Tab. 35. STIPAGROSTIS HIRTIGLUMA subsp. PATULA. 1, habit (×½); 2, ligule (×5); 3, glumes (×4½); 4, floret (×4½), all from *Goldsmith* 140/55.

having a ring of short hairs near the point and a ring of long hairs at the base of the lemma; column 7–10 mm. long, hairy towards the apex or twisted; central awn 5–7 cm. long, plumose all over or naked in the lower part and plumose towards the apex, with a naked excurrent apex; lateral awns up to 2·5 cm. long, very delicate, naked.

Subsp. **patula** (Hack.) de Winter in Kirkia, **3**: 134 (1963). TAB. **35**. Type: Mozambique, Boroma, *Menyharth* 601 (Z, holotype).

    Aristida hirtigluma var. *patula* Hack. in Denkschr. Kaiserl. Akad. Wiss. Wien, Math.-Naturw. Kl. **78**: 401 (1906). Type as above.

    Aristida gracilior Pilg. in Engl., Bot. Jahrb. **40**: 80 (1907).—Dinter in Fedde, Repert. **15**: 341 (1918).—Henrard, Crit. Rev. Aristida **1** in Meded. Rijks-Herb. **54**: 208 (1926); in Monogr. Aristida **1** in Meded. Rijks-Herb. **58**: 69, t. 15 (1929).—Schweickerdt in Bothalia, **4**: 123 (1941).—Chippindall in Meredith, Grasses & Pastures of S. Afr.: 303 (1955). Type from SW. Africa.

    Stipagrostis hirtigluma var. *patula* (Hack.) de Winter in Bothalia, **8**: 365, fig. 135, 136 & 137 (1965). Type as for *Stipagrostis hirtigluma* subsp. *patula*.

Perennial, densely caespitose. Panicle effuse and open, with long branches bearing spikelets on long pedicels. Glumes nearly of equal width, the inferior 9–11 mm. long, the superior 10–13 mm. long.

Botswana. N: Francistown-Shashi, 17.iv.1931, *Pole Evans* 3224 (K; SRGH). SW: Mabua Sehoba pan 116·8 km. S. of Tsani, 22.ii.1960, *de Winter* 7464 (K; SRGH). **Rhodesia.** S: Nuanetsi, Chikombedzi-Malipate road, 460 m., v.1955, *Davies* 1201 (K; SRGH). **Mozambique.** Tete, Boroma, *Menyharth* 601 (Z, holotype). SS: Alto Limpopo, between Mapai and R. Nuanetzi, 19.iv.1955, *Myre* 2085 (SRGH; LMA).

    Also in SW. Africa, the Cape Prov., the Transvaal and E. Africa. Usually on calcareous soils, limestone outcrops, on edges of pans, also on basalt stony ridges.

In his latest paper on S. African *Aristideae*, de Winter (1965) considered this taxon to be a variety of *S. hirtigluma*. Since it grows in less dry areas showing a more marked preference for calcareous soils, than typical forms do, and is characterised by a number of characters pointed out above, it is treated here as a subsp. as de Winter did in 1963. The typical subsp.—subsp. *hirtigluma*—has not been recorded from the Flora Zambesiaca area. It consists of annual forms, with a few basal leaves. Their panicles are longer than wide when fully exserted, having short branches with spikelets on short pedicels. The glumes are less than 10 mm. long and of unequal width, the inferior one being much broader than the superior. It is distributed in SW. Africa, S. Angola, N. Africa, Ethiopia, eastwards to Arabia and India. *S. hirtigluma* subsp. *patula* resembles *S. uniplumis*, but differs mainly in having a more strongly tuberculate upper part of the lemma gradually tapering into a column, a longer central awn with a naked excurrent apex and in the absence of the penicil at the apex of the column. The glumes in *S. hirtigluma* subsp. *patula* are hirtellous or ciliate, but in *S. uniplumis* usually glabrous.

XVII. CENTOTHECEAE Ridl.

Centotheceae Ridl., Mat. Fl. Malay. Penins. **3**: 122 (1907).

Inflorescence a loose or contracted panicle. Spikelets 1–many-flowered (if 1-flowered the fertile floret followed by several empty lemmas or rarely a few ♂ florets), ♀, all alike, falling entire or the rhachilla disarticulating below the florets. Glumes always shorter than the lemmas, 3–7-nerved, herbaceous to chartaceous, awnless. Lemmas 3–9-nerved, similar in textures to the glumes, awnless or shortly awned from the apex, sometimes with tubercle-based setae along the margins, otherwise glabrous. Palea as long as or slightly shorter than the corresponding lemmas, hyaline, 2-keeled. Lodicules 2, cuneate. Stamens 2 or 3. Styles 2, free or slightly connate at the base. Caryopsis loose between lemma and palea; embryo c. $\frac{1}{3}$ the length of the caryopsis; hilum basal, elliptic or ovate; starch grains simple or mixed. Chromosomes small, basic number 12.

    A tribe of c. 6 genera of tropical forest grasses. Leaf-laminae usually broad, tessellate. Ligule inconspicuous, membranous.

Glumes not persistent. Spikelet 1–5-flowered, falling entire at maturity
                                                         30. **Orthoclada**

Glumes persistent.   Spikelets 2–many-flowered, disarticulating between the florets
at maturity:
Lemmas shortly awned - - - - - - - - 32. **Bromuniola**
Lemmas awnless - - - - - - - - 31. **Megastachya**

30. ORTHOCLADA Beauv.

Orthoclada Beauv., Agrost.: 69, t. 14 fig. 9 (1812).

Spikelets pedicelled, laterally compressed, imbricate, awnless, 1–5-flowered, with all florets both fertile and ♀ except the uppermost one which is reduced; rhachilla glabrous or somewhat pubescent, not disarticulating between the florets; the internodes adnate to the base of the palea.   Glumes equal in size or the inferior one slightly shorter, keeled, 3–5-nerved, membranous to chartaceous, acute; the inferior narrowly to broadly lanceolate; the superior broadly lanceolate to lanceolate-oblong.   Lemmas 5–7-nerved, somewhat keeled or dorsally rounded, membranous to cartilaginous, elliptic or ovate-elliptic if expanded, acute or subobtuse, sometimes mucronulate; the lateral nerves usually faint.   Paleas slightly shorter than the corresponding lemmas, 2–4-nerved, 2-keeled, laterally compressed, the keels close together, scaberulous along the keels.   Lodicules 2, minute, cuneate, apex truncate, glabrous.   Stamens 2–3; anthers narrowly oblong.   Ovary glabrous; styles distinct;   stigmas plumose, laterally exserted.   Caryopsis free, laterally compressed, obliquely elliptic-oblong in outline; embryo c. ¼ the size of the caryopsis; hilum very minute, basal.   A perennial grass with erect or ascending culms. Ligule membranous, truncate, short.   Leaf-laminae with a short pseudopetiole; venation tessellate.   Inflorescences loose open panicles.
A genus of 2 described spp., 1 in tropical America, 1 in tropical Africa.

Orthoclada africana C. E. Hubb. in Hook., Ic. Pl. **35**: 3419 (1940).—W. D. Clayton, F.T.E.A. Gramineae: 163 (1970).   TAB. **36**.   Type: Zambia, Mwinilunga Distr., Luakera Falls, *Patterson* (K).

A robust perennial with a stout horizontally creeping rhizome.   Culms up to 180 cm. high, 6–9-noded, erect, stout, terete, striate, smooth, puberulous but finally glabrescent.   Leaf-sheaths striate, keeled, truncate and barbate at the mouth, rather tight when young, later somewhat loose and slipping off the culm.   Ligule minutely ciliolate along the upper edge.   Leaf-laminae (10)12–20(25) × (1·5)2–3·5 (4) cm., oblong-lanceolate to narrowly elliptic-oblong, at the rounded base constricted into a pseudo-petiole of 0·5–1·7 cm. in length, apex acute to acuminate, puberulous to glabrescent on both surfaces, smooth or somewhat scaberulous on the midrib beneath.   Panicle 20–35 × 8–20 cm., ovate or broadly ovate to elliptic in outline;   the branches mostly verticillate, spreading or obliquely ascending, slender, somewhat rigid, unbranched, scaberulous.   Spikelets 8–12 mm. long, 3–5-flowered, narrowly oblong, greenish.   Glumes 5–5·5 mm. long, obtuse to subacute, scaberulous on the keel, asperous on the flanks.   Lemmas 4·5–5 mm. long, decreasing in size towards the apex, obtuse to subacute, dorsally somewhat asperous.   Stamens 3; anthers 2–3 mm. long.   Caryopsis unknown.

Zambia. N: Kawambwa, 27.iii.1962, *Lawton* 838 (SRGH). W: Mwinilunga Distr., Luakera Falls, 25.i.1938, *Milne-Redhead* 4333 (BM; K; PRE).
Also in Katanga and Tanzania.   In forest shade, often near rivers, also in open patches in thickets.

31. MEGASTACHYA Beauv.

Megastachya Beauv., Agrost.: 74 (1812).

Spikelets pedicelled, solitary, 8–18(31)-flowered, laterally compressed, loosely imbricate.   Rhachilla easily disarticulating above the glumes and between the florets.   Florets all fertile and ♀ except for the terminal one which is usually reduced. Glumes slightly unequal, both shorter than the following floret, 3–(rarely 5) nerved, subpersistent, chartaceous, the middle nerve sometimes excurrent into a short

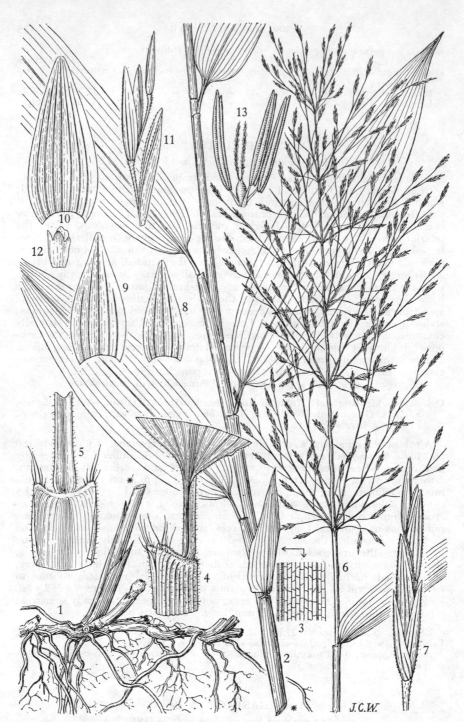

Tab. 36. ORTHOCLADA AFRICANA. 1, rhizome ($\times \frac{2}{3}$); 2, leafy culm ($\times \frac{2}{3}$); 3, part of underside of leaf-lamina showing tessellate venation ($\times 3\frac{1}{2}$); 4, false petiole ($\times 3$); 5, ligule ($\times 2$); 6, inflorescence ($\times \frac{2}{3}$); 7, spikelet ($\times 6$); 8, inferior glume ($\times 8$); 9, superior glume ($\times 8$); 10, lemma ($\times 8$); 11, palea and sterile lemmas ($\times 8$); 12, lodicules ($\times 12$); 13, sexual organs ($\times 8$), all from *Milne-Redhead* 4333.

awn, keeled, glabrous and smooth except for the keels which are scaberulous towards the apex. Lemmas 7-nerved, chartaceous, with the apex shortly 2-fid or rarely entire and the middle nerve running out into a short mucro between the lobes, glabrous, scaberulous along the keel, otherwise smooth or very rarely the flanks somewhat asperous towards the apex; the lateral nerves equally distant, evanescent below the apex, rather inconspicuous in reflected light. Paleas almost as long as the corresponding lemmas, sometime protruding, 2-keeled, membranous, apex truncate or shortly emarginate, the inflexed margins very narrow. The keels scabrous or shortly and stiffly ciliolate. Lodicules very minute, cuneate, apex obliquely truncate. Stamens 2 or 3. Ovary glabrous; styles distinct; stigmas plumose. Caryopsis elliptic in outline, regularly triangular in cross-section, subacute at both ends, black when mature; embryo $\pm \frac{1}{3}$ the length of the caryopsis, invisible; hilum very minute, basal. Annual or short lived perennial. Ligule a short membrane. Leaf-laminae with tessellate nervation. Inflorescences open or (rarely) somewhat contracted panicles.

A monospecific genus. Known from west tropical Africa, the Congo, Angola Tanzania, Madagascar, Mozambique, Zambia, Zululand and Natal.

Megastachya mucronata (Poir.) Beauv., Agrost.: 74 (1812).—W. D. Clayton in F.T.E.A., Gramineae: 161 (1970). TAB. **37**. Type from ? Tropical Africa.
   Poa mucronata Poir. in Encycl. Méth., Bot. **5**: 91 (1804).—Beauv., Fl. Owar. Ben.: 5, t. 4 (1805). Type as above.
   Centotheca mucronata (Poir.) Kuntze, Rev. Gen. Pl.: 765 (1891).—Hutch. & Dalz., F.W.T.A. **2**: 505 (1931). Type as for *Megastachya mucronata*.
   Centotheca maxima Peter, Fl. Deutsch Ost-Afr. **1**: 340, Anh. 113 (1936). Type from Tanzania.

An annual or short-lived perennial. Culms 30–110 cm. high, erect or more often decumbent, rooting and emitting shoots from the lower nodes as well, 2–5-noded, terete, finely striate, glabrous and smooth. Leaves usually densely crowded around the base, distant along the culms; sheaths somewhat tight when young, later lax and slipping off the culm, keeled, striate, glabrous and smooth; lamina (5)9–10(15) × (0·8)1·25–2(3) cm., lanceolate, broadly lanceolate to narrowly elliptic, tapering to a fine point, base auriculate or broadly rounded, inconspicuously puberulous on both surfaces but soon becoming glabrous, scaberulous along the edges, otherwise smooth. Panicle 12–35 × 5–25 cm., usually lax and open, broadly elliptic or obovate in outline, erect, often partly enclosed in the sheath of the uppermost leaf; main axis obtusely angular or sulcate, asperulous along the ridges, glabrous; branches up to 18 cm. long, spreading or obliquely ascending, usually rather lax, branched from near the base, the subdivisions filiform, asperulous. Pedicels up to 15 mm. long, filiform, asperulous. Spikelets 9–21 × 2–3 (rarely–4) mm., oblong to linear-oblong. Glumes ovate when flattened; the inferior 1·5–2·3 mm. long; the superior 1·8–3 mm. long. Lemma 2·2–3·5 mm. long, broadly ovate to very broadly elliptic-ovate. Anthers 0·6–1·4 mm. long. Caryopsis c. 1·2 × 0·6 mm.

Zambia. B: Mongu, 10.xi.1959, *Drummond & Cookson* 6297 (BM; SRGH). N: Kasama, near Chambeshi Pontoon, 1190 m., 7.iv.1961, *Phipps & Vesey-FitzGerald* 2989 (BM; SRGH). W: Kasempa, Busanga Plain, 18.viii.1961, *Mitchell* 8/37 (K; LISC; SRGH). **Rhodesia.** E: Melsetter, Ngorime Reserve, Lusitu R. beneath Hayfield B., c. 1200, 29.xi.1967, *Simon & Ngoni* 1372 (K; SRGH). **Malawi.** N: Nkhata Bay, Chombe Estate, 10.ix.1955, *Jackson* 1752 (K; LISC). **Mozambique.** N: Macondes, 37 km. between Maeda and Mocímboa do Rovuma, c. 800 m., 15.iv.1964, *Torre & Paiva* 12009 (LISC). Z: Quelimane, Maganja da Costa, 31.vii.1943, *Torre* 5718 (LISC; PRE). MS: Garuso Forest, iv.1935, *Gilliland* 1855 (BM; K). SS: Inharrime, Nhacoongo near to Sr. João Duarte de Silva, 4.iv.1959, *Barbosa & Lemos* 8489 (K; LISC; LMA; PRE). LM: Between Lourenço Marques and Costa do Sol, 1.iv.1948, *Schweickerdt* 1909 (BM; K; PRE).

Occurs in western tropical Africa, the Congo, Angola, Tanzania, Zululand, Natal and Madagascar. Frequently in the shade of coastal bush, also in woodland, favouring the shade of trees, usually on heavy loamy soil and in swamp thicket.

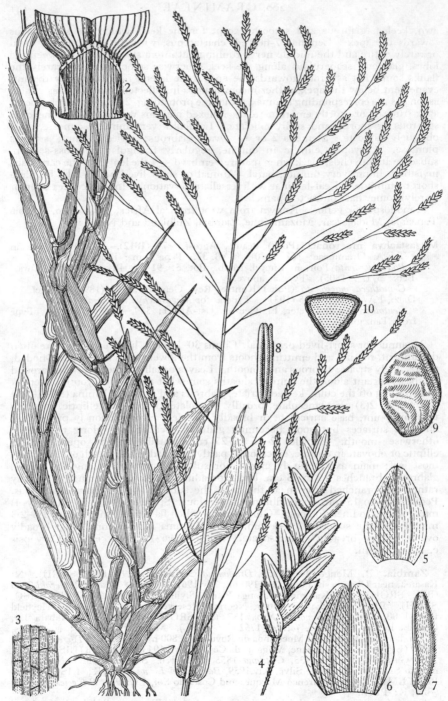

Tab. 37. MEGASTACHYA MUCRONATA. 1, habit (× ⅔); 2, mouth of leaf-sheaths and basal part of lamina showing ligule (× 3); 3, part of the lower surface of leaf-lamina showing tessellate venation (× 4); 4, spikelet (× 5); 5, superior glume (× 12); 6, lemma (× 14); 7, palea (× 14); 8, anther (× 14); 9, caryopsis (× 22); 10, transverse section of caryopsis (× 22), all from *Phipps & Vesey-FitzGerald* 2986.

32. BROMUNIOLA Stapf & Hubb.

Bromuniola Stapf & Hubb. in Kew Bull. **1926**: 366 (1926).

Spikelets long-pedicelled, solitary, laterally compressed, loosely imbricate, 3–7-flowered, awned.   Florete all fertile, ♀ except for the reduced terminal one. Rhachilla disarticulating above the glumes and beneath the florets.   Glumes subequal, persistent, keeled, membranous, acute, prominently 5-nerved, with transverse veinlets between the nerves, scaberulous along the keels, glabrous, often mucronate.   Lemmas similar to the glumes, 7-nerved, acute and shortly awned, keeled, glabrous and smooth except for the scaberulous keels and awns, membranous with hyaline margins, glabrous, the lowermost sterile.   Paleas slightly shorter than the lemmas, 2-keeled, hyaline, compressed, apex obtuse, base bulging out from the lemma, ciliate along the keels.   Lodicules 2, broadly cuneate.   Stamens 3.   Ovary elliptic, glabrous; styles distinct, short; stigmas plumose, laterally exserted. Caryopsis free, obliquely ovate in outline, laterally compressed; embryo very small; hilum basal.   Inflorescences of large panicles with spreading lax delicate branches.

A monospecific genus confined to tropical Africa.

Bromuniola gossweileri Stapf & Hubb. in Kew Bull. **1926**: 366 (1926).—W. D. Clayton in F.T.E.A., Gramineae: 161 (1970).   TAB. **38**.   Type from Angola.

A caespitose perennial from a creeping rhizome.   Culms up to 85 cm. high 5–7-noded, erect or geniculate-ascending, often straggling, rather weak, terete or somewhat compressed, striate, glabrous, smooth.   Leaf-sheaths striate, tight when young, later lax and slipping off the culm, smooth, glabrous.   Leaf-laminae (5)8–18(22) × 1–2·5 cm., constricted at the base, lanceolate or lanceolate-elliptic (usually broadest about the middle), tapering to an acute apex, with tessellate veins, expanded, glabrous or finely puberulous on both surfaces but usually soon glabrescent, sometimes softly pilose towards the base, asperulous along the margins, vivid green.   Panicle 10–20 × 12–25 cm., circular or broadly elliptic in outline, erect, rigid; main axis slender, obtusely angular, scaberulous, glabrous; branches up to 18 cm. long, solitary, flexuous, simple or few branched; the divisions filiform, asperulous.   Pedicels slender, filiform, flexuous, asperulous, up to 5 cm. long.   Spikelets 1–2·3 × 0·4–0·5 cm., oblong, often somewhat broader at the apex at maturity, bright green.   Glumes ovate-lanceolate to ovate; the inferior 4–5·5 mm. long; the superior 6–8 mm. long.   Lemmas 8–11 mm. long (excluding the awn), ovate; the awn 2·5–4 mm. long.   Paleas c. 6 mm. long.   Anthers 2–3 mm. long, linear.   Caryopsis 2·8–3·1 mm. long, reddish-brown with a green pericap when young.

Zambia.   W: Chichele, near Ndola, 12.v.1950, *Jackson* 47 (BM; K).
Also in the Congo Republic, Angola and Tanzania.   In forest glades.

XVIII. PAPPOPHOREAE Kunth

Pappophoreae Kunth, Rev. Gram.: 82 (1829).

Inflorescence a panicle, usually dense, open or contracted.   Spikelets all alike in shape, 2–many-flowered; lower florets ♀, the upper reduced to a varying degree; rhachilla disarticulating above the glumes.   Glumes 2, persistent, subequal, membranous, slightly shorter than or as long as the spikelets, dorsally rounded, distinctly (1)3–9-nerved, with the nerves often rather irregular.   Lemma firmly membranous, deeply lobed, dorsally rounded, 9–many-nerved with the nerves conspicuously produced into 5–many awns; lobes usually hyaline.   Palea 2-nerved, usually as long as the body of the corresponding lemma.   Lodicules 2.   Stamens 3. Ovary glabrous; styles 2; stigmas 2, plumose.   Caryopsis with the punctiform hilum basal; starch grains compound.   Ligule a fringe of hairs.   Chromosomes small, basic number 10.

A tribe of 5 genera, widely distributed throughout the tropics and subtropics of both hemispheres.

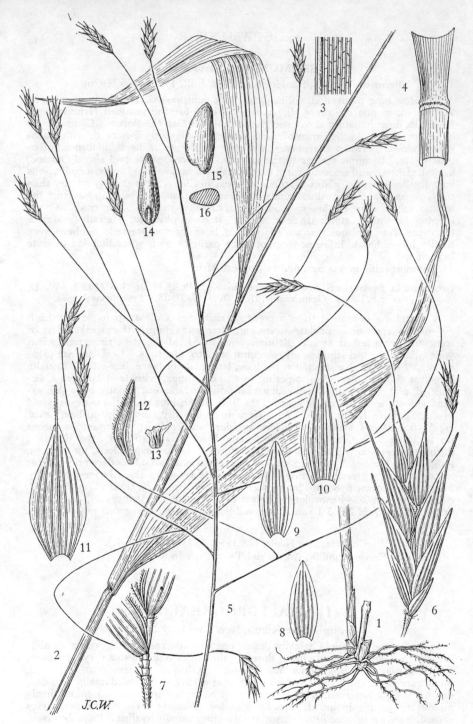

Tab. 38. BROMUNIOLA GOSSWEILERI. 1, base of plant (× ⅔); *Vesey-FitzGerald* 2594; 2, part of culm (× ⅔); 3, part of lower surface of leaf-lamina showing venation (× 4); 4, ligule (× 2); 5, inflorescence (× ⅔); 6, spikelet (× 3); 7, rhachilla (× 4); 8, inferior glume (× 4); 9, superior glume (× 4); 10, lowest (sterile) lemma (× 4); 11, lemma (× 4); 12, palea (× 4); 13, lodicule (× 4); 14, caryopsis, ventral view (× 6); 15, same in lateral view (× 6); 16, same, transverse section (× 6), all from *Jackson* 47.

Lemma 9-awned, with the awns emerging from the apex of 9 hyaline lobes
33. **Enneapogon**
Lemma 5-awned, with the awns alternating with 4 hyaline lobes -   - 34. **Schmidtia**

33. ENNEAPOGON Desv. ex Beauv.

Enneapogon Desv. ex Beauv., Agrost.: 81 (1812).

Spikelets awned, solitary, slightly laterally compressed, 3–6-flowered; rhachilla disarticulating above the glumes and below the fertile florets, terete; florets 3 in African spp.; the lowermost ♀; the 2nd ♂ or sterile; the terminal reduced, often only a tuft of awns. Glumes 2, persistent, 3–5 (rarely sub-7)-nerved, equal to slightly unequal, membranous, dorsally rounded or inconspicuously keeled; the superior one about as long as the inferior lemma (excluding the awns). Lemma of the inferior (♀) floret 9-nerved, 9-awned, chartaceous, rather firm, broad, dorsally rounded, densely pilose (villous) at least in the lower ½; awns equal or subequal, subulate, straight, scaberulous or plumose, (2)3–5 times as long as the body of the lemma; callus minute. Palea 2-nerved, 2-keeled, slightly longer than the lemma body, hyaline, often ciliolate along the keels. Second lemma similar but only ½ as long as the first. Lodicules 2, minute, glabrous, fleshy. Stamens 3. Ovary glabrous; styles distinct, short; stigmas plumose, laterally exserted. Caryopsis oblong to broadly elliptic in outline, dorsally somewhat compressed or shallowly grooved; embryo rather large, about ¾ the length of the caryopsis, hilum punctiform, basal. Predominantly perennials, rarely annual grasses, of various habits. Ligule reduced to a fringe of hairs. Inflorescence usually a contracted often spike-like panicle.

A genus of c. 30 spp., 19 of which occur in Australia; 1 sp. is known from North and Central America; 7 spp. are present in Africa, the rest occur in Asia.

Anthers 0·3–0·5 (rarely –0·7) mm. long; small, densely caespitose perennials with the lowermost leaf-sheath usually splitting into persistent fibres, often conspicuously stoloniferous; cleistogamous spikelets often produced within the basal and nodal leaf-sheaths -   -   -   -   -   -   -   -   -   -   -   - 1. *desvauxii*
Anthers (0·8)1·2–2·7 mm. long; annuals or perennials, if perennials then the lowermost leaf-sheath not splitting into persistent fibres, rarely stoloniferous; leaf-sheaths always without cleistogamous spikelets:
  Annuals or rarely short-lived perennials, shortly and densely glandular pilose; culms fairly stout, usually geniculately ascending from a loosely caespitose never bulbous base; leaf-laminae expanded or rarely with the margin involute, 3–10 mm. wide
2. *cenchroides*
  Perennials, often growing with a strong horizontal or oblique rhizome, loosely pubescent to almost glabrous; culms rather slender, wiry, usually erect, rarely abruptly ascending from a very shortly decumbent often bulbous base; leaf-laminae almost always involute, setaceous, tapering to a fine curled point, up to 3 mm. wide (when expanded):
    Anthers (0·8) 1·2–2(2·2) mm. long; culms with a sub-bulbous base, profusely branched above, loosely pubescent; inflorescence a densely contracted spike-like panicle -   -   -   -   -   -   -   -   -   -   - 3. *scoparius*
    Anthers 2·3–2·7 mm. long; culms not or only inconspicuously bulbous at the base, simple (or very rarely branched below ?), inconspicuously pubescent to almost glabrous; inflorescence a loosely contracted or somewhat open panicle; apices of glumes (and often the tips of the awns of the lemmas) usually brown or rust-coloured -   -   -   -   -   -   -   -   -   - 4. *pretoriensis*

1. **Enneapogon desvauxii** Beauv., Ess. Agrost.: 82, t. 16, f. 11 (1812).—W. D. Clayton in F.T.E.A., Gramineae: 167 (1970). TAB **39** fig. B. Type ?from S. America.
   Pappophorum brachystachyum Jaub. & Spach, Ill. Pl. Or. **4**: 34, t. 324 (1851). Type as above.
   Pappophorum bulbosum Fig. & De Not. in Mem. Acad. Sci. Torino, **2**, 12: 254 (1852). Type from Arabia.
   Pappophorum figarianum Fig. & De Not., loc. cit. Type from the Sudan.
   Pappophorum vincentianum J. A. Schmidt, Beitr. Fl. Cap. Verd. Ins.: 144 (1852). Type from the Cape Verde Is.
   Pappophorum arabicum Hochst. ex Steud., Syn. Pl. Glum. **1**: 199 (1854). Type from Arabia.
   Pappophorum nanum Steud., Syn. Pl. Glum. **1**: 200 (1854) *nom. illegit.* based on *P. vincentianum* J. A. Schmidt.

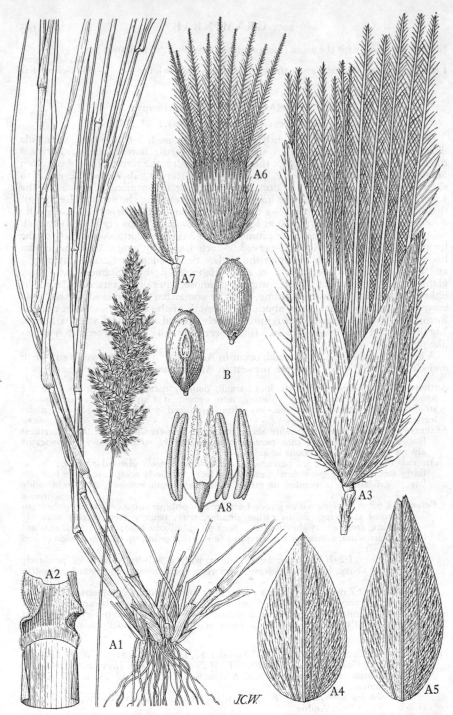

Tab. 39. A.—ENNEAPOGON CENCHROIDES. A1, habit (× ⅔); A2, junction of leaf-sheath and lamina, showing ligule (× 3); A3, spikelet in lateral view (× 30); A4, inferior glume (× 12); A5, superior glume (× 12); A6, lemma (× 12); A7, palea, showing also lodicule and extension of rhachilla with a reduced floret (× 12); A8; sexual organs (× 14) all from *Gomes e Sousa* 3371. B.—ENNEAPOGON DESVAUXII, caryopsis, dorsal and ventral view respectively (× 14) *Drummond* 5949.

Pappophorum senegalense Steud., tom. cit.: 199 (1854). Type from Senegal.
Enneapogon pusillus Rendle, Cat. Afr. Pl. Welw. **2**, 1: 229 (1899). Type from Angola.
Enneapogon brachystachyus (Jaub. & Spach) Stapf in Harv. & Sond., F.C. **7**: 654 (1900).—Bremek. & Oberm. in Ann. Transv. Mus. **16**: 405 (1935).—Chippindall in Meredith, Grasses & Pastures of S. Afr.: 237 (1955).—Bogdan, Rev. List Kenya Grass.: 16 (1958).—Bor, Grass. B.C.I. & P.: 608 (1960).—Napper, Grass. Tangan.: 21 (1965).—Renvoize in Kew Bull. **22**: 396 (1968).—Launert in Merxm., Prodr. Fl. SW. Afr. **160**: 78 (1970). Type from Arabia.
Enneapogon brachystachyus var. *macranthera* Stapf in Harv. & Sond., F.C. **7**: 655 (1900).—Chippindall in Meredith, Grasses & Pastures of S. Afr.: 238 (1955). Type from S. Africa.
Pappophorum pusillum (Rendle) K. Schum. in Just's Bot. Jahresber. **27**: 458 (1901). Type as for *Enneapogon pusillus*.

A compact caespitose often stoloniferous perennial, usually densely glandular-pubescent all over. Culms 5–15 (rarely to 40) cm. tall, 2–5-noded, geniculately ascending, sometimes decumbent, often with a bulbous thickening at the base, simple or branched, rather slender, asperulous. Leaves mostly confined to the base, forming dense tufts; sheaths tight at first later somewhat loose, finely to coarsely striate, the oldest long-persistent and splitting into fibres which form dense cushions at the base of the culm. Leaf-laminae 2·5–7·5(12) × 0·3–0·5 cm., filiform, almost always convolute, tapering to a very fine point, flexible, asperulous. Panicle 1·5–5(7·5) cm. long, spike-like, narrowly cylindrical or rarely ovate-oblong in outline, rather dense, light to dark grey. Spikelets 3–5·5 mm. long, 3-flowered (see note below), crowded. Glumes subequal, oblong, scantily pilose, with the apex obtuse to slightly emarginate, light to dark grey; the inferior 2·3–5 mm. long, 3–7-nerved; the superior 2·8–5·5 mm. long, 3–5-nerved. Fertile lemma (the inferior) 1·5–2 mm. long (excluding the awns), dorsally shortly villous; awns 2–4 mm. long, shortly plumose to or beyond the middle. Anthers 0·3–0·5 (rarely 0·7) mm. long. Caryopsis c. 1 mm. long.

Botswana. N: NE. corner of Makarikari Pan, 16.i.1959, *West* 3863 (BM; SRGH). SW: 9·6 km. SW. of Takatshwane on road to Lehututu, 21.ii.1960, *de Winter* 7420 (K; PRE; SRGH). SE: Khomodimo Pan, 206 miles NW. of Molepolole, 18.vi.1955, *Story* 4926 (K; PRE). **Rhodesia.** S: Beitbridge, Shashi-Limpopo confluence, 22.iii.1959, *Drummond* 5949 (BM; SRGH).
Distributed in North Africa from the Cape Verde Is. to Egypt, through Arabia to India and China, in tropical Africa from Chad and Somalia southwards to the Cape Prov., also occurring in SW. Africa and southern Angola. Growing in open grassland, in limestone pans, on grassy hillsides and stony outcrops, also on thin soil overlapping basalt rocks, often in overgrazed areas.

This sp. often produces cleistogamous spikelets, which are usually enclosed within the bulb-like enlarged basal leaf-sheaths. The caryopsis of these spikelets is conspicuously larger than that of normal spikelets, and remain enclosed within the sheaths where it finally germinates. Occasionally small panicles, which produce normal spikelets, are also produced within basal or nodal leaf-sheaths.

2. **Enneapogon cenchroides** (Licht.) C. E. Hubb. in Kew Bull. **1934**: 119 (1934).—Eggeling, Annot. List Grass. Uganda: 17 (1947).—Sturgeon in Rhod. Agr. Journ. **51**: 225 (1954).—Chippindall in Meredith, Grasses & Pastures of S. Afr.: 236, 238 (1955).—F. W. Andr., Fl. Pl. Sudan, **3**: 448 (1956).—Bogdan, Rev. List Kenya Grass.: 16 (1958).—Harker & Napper, Illustr. Guide Grass. Uganda: 28 (1960).—Bor, Grass. B.C.I. & P.: 608 (1960).—Napper, Grass. Tangan.: 21 (1965).—Renvoize in Kew. Bull. **22**: 397 (1968).—W. D. Clayton in F.T.E.A., Gramineae: 169 (1970).—Launert in Merxm., Prodr. Fl. SW. Afr. **160**: 78 (1970). TAB. **39**, fig. A. Type from S. Africa.
Pappophorum cenchroides Licht. in Roem. & Schult., Syst. Veg. **2**: 616 (1817). Type as above.
Enneapogon mollis Lehm., Nov. Stirp. Pugill., **3**: 40 (1831).—Stapf in Harv. & Sond., F.C. **7**: 654, 655 (1900).—Bremek. & Oberm. in Ann. Transv. Mus. **16**: 405 (1935).—Sturgeon in Rhod. Agric. Journ. **51**: 225 (1954). Type from S. Africa.
Pappophorum molle (Lehm.) Kunth, Enum. Pl. **1**: 255 (1833). Type as above.
Pappophorum abyssinicum Hochst. in Flora, **38**: 202 (1855). Type from Ethiopia.
Pappophorum robustum Hook. f., Fl. Brit. Ind. **7**: 302 (1896). Type from India.
Enneapogon abyssinicus (Hochst.) Rendle, Cat. Afr. Pl. Welw. **2**, 1: 229 (1899). Type as for *Pappophorum abyssinicum*.

A loosely caespitose annual or rarely shortly lived perennial, densely glandular-pubescent all over. Culms 15–100 cm. tall, 2–5-noded, rather stout, geniculately ascending, seldom erect, sometimes decumbent, simple or branched below. Leaf-sheaths relatively tight, usually slightly shorter than the internodes, striate, smooth or somewhat asperulous towards the mouth. Leaf-laminae 3–25 × (0·1)0·3–0·7(1) cm., linear to lanceolate-linear, long-tapering to a fine point, expanded or convolute (often only towards the apex), rigid to subflaccid, scaberulous on the upper surface and along the margins. Panicle 3–15(20) cm. long, spike-like, usually dense and contracted, rarely somewhat open below, compact, rarely interrupted. Spikelets 3–5 mm. long, usually crowded, 3-flowered. Glumes slightly unequal, light to dark grey or grey-green; the inferior 2·8–4(5·1) mm. long, 5–7-nerved, ovate; the superior 3·2–5·5(6·8) mm. long, 3-nerved, oblong, often with the apex somewhat truncate. Fertile lemma (the inferior) 1·5–2 mm. long (excluding the awn), dorsally shortly villous; awns 3–4.25 mm. long, plumose up to or beyond the middle. Palea 2–2·25 mm. long, with the keels ciliolate. Anthers 0·8–1·8(2·3) mm. long.

Botswana. N: Lake Ngami near Sehitwa, 24.iii.1961, *Vesey-FitzGerald* 3327 (BM; SRGH). SW: Damara Pan, 20.iv.1930, *Van Son* 28630 (BM; COI; K; SRGH). SE: 246·4 km. NW. of Molepolole, 16.vi.1955, *Story* 4907 (COI; K; SRGH). **Rhodesia.** N: Sebungwe, near Binga, 451 m., 6.xi.1958, *Phipps* 1369 (BM; SRGH). W: Wankie, 9.v.1955, *Plowes* 1805 (K; SRGH). E: Chipinga, Sabi Valley Experiment Station, 1959, *Soane* 86 (BM; SRGH). S: Beitbridge, Masena, 7.vi.1962, *West* 4113 (BM; SRGH). **Mozambique.** SS: Caniçado, Regulado Chirunzo, 18.i.1947, *Pedrogão* 333 COI; K; LMA). LM: Lourenço Marques, Marracuene, 26.xi.1940, *Hornby* 3103a (LISC).
In Uganda, Kenya, Tanzania, Angola, SW. Africa, the northern and NW. Cape Prov., Transvaal, Orange Free State, Ascension I. and through Arabia to India. Widely distributed in woodland, open grassland, cultivated and disturbed areas, in dried up streams etc., usually on poor sandy soil, often on rocky hillsides.

3. **Enneapogon scoparius** Stapf in Harv. & Sond., F.C. **7**: 656 (1900).—Chippindall in Meredith, Grasses & Pastures of S. Afr.: 235, 237 (1955).—Renvoize in Kew Bull. **22**: 397 (1968).—Launert in Merxm., Prodr. Fl. SW. Afr. **160**: 79 (1970). Type from S. Africa.
  Pappophorum setifolium Hochst. in Act. Hort. Petrop. **1**: 28 (1871) *nom. nud.*
  Pappophorum scoparium (Stapf) Chiov. in Ann. Ist. Bot. Roma, **8**: 358 (1908). Type as for *Enneapogon scoparius.*
  Pappophorum filifolium Pilg. in Engl., Bot. Jahrb. **51**: 419 (1914). Type from SW. Africa.
  Enneapogon filifolius (Pilg.) Stapf ex Garabedian in Ann. S. Afr. Mus. **16**: 424 (1925). Type as above.

A perennial, with a short horizontal rhizome, usually inconspicuously pubescent all over, often appearing glabrous. Culm 20–80 cm. tall, few–6-noded, very slender, wiry, arising from a bulbous base, usually branched a few cm. above the base, usually erect, rarely geniculately ascending. Leaf-sheaths rather tight, finely striate, the lowermost long-persistent. Leaf-laminae 5–20(25) × 0·1–0·3 cm., almost always convolute, filiform, very rarely expanded, tapering to a fine point, somewhat rigid, obliquely ascending to erect, usually smooth. Panicle 1·5–12 cm. long, dense, spike-like, compact, rarely interrupted, light to dark grey. Spikelets 3·5–4·5 mm. long, 3-flowered, crowded. Glumes slightly unequal, scattered-pilose, subacute; the inferior 2·5–5 mm. long, 3–7-nerved, ovate; the superior 3·4–7 mm. long, 3–6-nerved, oblong. Fertile lemma (the inferior) c. 2 mm. long (excluding the awns), dorsally villous; awns 2–3·5 mm. long, plumose beyond the middle. Anthers (0·8)1·2–2·2 mm. long.

Botswana. N: NE. corner of Makarikari Pan, 16.i.1959, *West* 3860 (BM; SRGH). SE: 53 km. NW. of Serowe, 25.iii.1965, *Wild & Drummond* 7310 (BM; K; LISC; SRGH). **Rhodesia.** E: Chipinga, 385 m., 19.i.1957, *Phipps* 11 (COI; LISC; SRGH). S: Ndanga, 21.vii.1959, *West* 3971 (BM; SRGH). **Mozambique.** SS: Limpopo R. Limpopo between Mapai and R. Nuanetsi, 19.iv.1955, *Myre* 2084 (SRGH; LMA).
Widely distributed throughout southern Angola, SW. Africa, the NW. Cape Prov., Transvaal, Orange Free State, Natal, also in Ethiopia. Mostly in *Colophospermum mopane* woodland, but also in scrub and open grassland, growing in arid conditions, amongst limestone pebbles, on basaltic soil, also on sandy hill-slopes and along roads.

4. **Enneapogon pretoriensis** Stent in Bothalia, **1**: 174, pl. 3 (1922).—Chippindall in
Meredith, Grasses & Pastures of S. Afr.: 235, 236 (1955).—Renvoize in Kew Bull.
22: 399 (1968). Type from S. Africa.

A caespitose perennial, with a short oblique or almost vertical rhizome, very
inconspicuously pubescent to almost glabrous. Culms 30–75 cm. tall, 2–5-noded,
slender but less wiry than in the preceding sp., simple (!), erect or sometimes
ascending from a shortly geniculate base. Leaf-sheaths tight at first, later somewhat loose but always firm, the lowermost long-persistent, pale. Leaf-laminae
5–25 × 0·2–0·4 cm., usually with the margins involute, rarely expanded (sometimes towards the base only), tapering to a fine pungent point, somewhat rigid.
Panicle up to 15 cm. long, ovate-oblong to oblong in outline, loose and somewhat open, rarely dense but never spike-like; branches up to 5 cm. long, obliquely
ascending to erect, often appressed to the rhachis. Spikelets 5·5–7 mm. long,
3-flowered, less crowded than in the other spp. Glumes slightly unequal, with the
apices usually bright orange, brown or rubiginous; the inferior 4–6 mm. long,
7–9-nerved; the superior 4·4–6·6 mm. long, 5-nerved. Fertile lemma (the inferior)
2·5–3 mm. long (excluding the awns), dorsally villous; awns 2–4·25 mm. long,
plumose beyond the middle towards the apices, often tinged with orange or brown.
Anthers 2·3–2·7 mm. long.

Botswana. SE: Kanye Hill, 22.iv.1958, *de Beer* 690 (SRGH).
Also in the Transvaal and the Orange Free State. On rocky hills in grassland or
open woodland. According to Chippindall (loc. cit.) it is usually found on slopes with a
northern aspect.

34. SCHMIDTIA Steud. ex J. A. Schmidt

Schmidtia Steud. ex J. A. Schmidt, Beitr. Fl. Cap. Verd. Ins.: 144 (1852)
nom. conserv.

Spikelets shortly pedicelled to subsessile, slightly laterally compressed, 4–6(–9)-
flowered. Florets ♀ except the uppermost 1 or 2 which are usually sterile and reduced to a varying degree; rhachilla disarticulating above the glumes and (tardily)
between the florets. Glumes 2, slightly different in length, persistent, (7)9–11(–14)-
nerved, membranous, acute or subobtuse, not or inconspicuously keeled. Lemma
dorsally rounded, subcoriaceous, dorsally long-villous towards the base, usually
scabrid, 6-lobed to nearly the middle with 5 scabrid awns between the membranous
lobes; callus minute, pointed, bearded. Palea always longer than the solid body
of the corresponding lemma, oblong or elliptic-oblong or broadly lanceolate,
2-keeled, with narrow flaps, stiffly ciliate along the keels and very often with inter-
spersed gland-tipped hairs. Lodicules 2, minute, cuneate-truncate, or often with
the apex shallowly 2-lobed and with gland-tipped hairs, carnose. Stamens 3.
Ovary glabrous; styles distinct, slender; stigmas plumose, laterally exserted.
Caryopsis oblong in outline; embryo ½–nearly ¾ the length of the caryopsis;
hilum minute, punctiform to oblong, sub-basal. Perennials or annuals of varying
habit, culms decumbent or more rarely erect, usually glandular-pilose and viscous.
Ligule reduced to a rim of stiff sericeous hairs. Inflorescence a panicle, loose
or contracted.

A genus of 2 spp., confined to Africa including the Cape Verde Is.

Plant conspicuously perennial, with a short creeping rhizome and usually long creeping
surface stolons. Culms usually bulbous at the base, often branched from the lower
nodes and with dense bundles of intravaginal innovation shoots, thus being of a
suffrutescent habit, often woody. Leaf-laminae 0·2–0·75 (rarely more) cm. wide,
tapering to a fine sometimes somewhat flexuous point, pubescent and often slightly
viscous from gland-tipped hairs (also the culms!), rarely glabrous. Keels of the paleas
either with or without a few gland-tipped hairs only towards the apex   1. *pappophoroides*
Plant annual, always without rhizome or stolons. Culms usually ascending from a
prostrate base and rooting at the lower nodes, without intravaginal innovation-shoots,
usually densely pubescent and viscous from gland-tipped hairs. Leaf-laminae (0·3)
0·7–1·75 cm. wide, abruptly tapering to a sharp point, viscous and hairy as the culms.
Keels of the paleas always with gland-tipped hairs, these usually conspicuously over-
topping the cilia and scattered over the whole length of the keels, more rarely confined
to the upper ⅓ -   -   -   -   -   -   -   -   -   2. *kalahariensis*

1. **Schmidtia pappophoroides** Steud. ex J. A. Schmidt, Beitr. Fl. Cap. Verd. Ins.: 145 (1852); Syn. Pl. Glum. **1**: 199 (1854).—C. E. Hubb. in F.W.T.A. **2**: 510 (1936).—F. W. Andr., Fl. Pl. Sudan, **3**: 529 (1956).—Bogdan, Rev. List Kenya Grass.: 16 (1958).—Launert in Bol. Soc. Brot., Sér. 2, **39**: 308 (1965); in Merxm., Prodr. Fl. SW. Afr. **160**: 169 (1970).—W. D. Clayton, F.T.E.A., Gramineae: 165 (1970). TAB. **40**. Type from the Cape Verde Is.

    Schmidtia quinqueseta Benth. ex Ficalho & Hiern in Trans. Linn. Soc., Bot., Sér. 2, **2**: 31 (1881). Type from Angola.

    Antoschmidtia quinqueseta (Benth.) Boiss., Fl. Orient. **5**: 559 (1884). Type as above.

    Schmidtia bulbosa Stapf in Harv. & Sond., F.C. **7**: 658 (1900).—R.E. Fr., Wiss. Ergebn. Schwed. Rhod.-Kong.-Exped. **1**: 210 (1916).—C. E. Hubbard in Mem. N. Y. Bot. Gard. **9**, **1**: 104 (1954).—Sturgeon in Rhod. Agric. Journ. **51**: 226 (1954).—Chippindall in Meredith, Grasses & Pastures of S. Afr.: 232 (1955).—Jackson & Wiehe, Annot. Check List Nyasal. Grass.: 58 (1958).—Harker & Napper, Ill. Guide Grass. Uganda: 52 (1960).—Napper, Grass. Tangan.: 21 (1965). Syntypes from S. Africa.

    Schmidtia glabra Pilg. in Engl., Bot. Jahrb. **43**: 386 (1909). Type from S. Africa.

    Antoschmidtia bulbosa (Stapf) Peter in Fedde, Repert. Beih. **40**: 307 (1931). non Bremek. & Oberm. (1935). Type as for *Schmidtia bulbosa*.

Perennial, with a short creeping rhizome, often with long surface stolons. Culms up to 100 cm. tall (usually much less!), 3- to many-noded, erect or geniculately ascending, almost always bulbous-like thickened at the base, simple or more often branched from the lower nodes, often with crowded intra-vaginal innovation-shoots and then suffrutescent in appearance, sometimes woody, finely striate, pubescent or sometimes glabrous, often somewhat viscous with gland-tipped hairs; the lower internodes usually short, the uppermost long exserted and slender. Leaf-sheaths tight when young, later loose and slipping off the culm; the lowermost scale-like, subcoriaceous, densely packed and often sericeous or villous. Leaf-laminae 5–16 × 0·2–0·7 (rarely more) cm., linear or linear-lanceolate, tapering to a setaceous point, sometimes somewhat curling towards the apex, involute or flat, glaucous, pubescent on both surfaces and sometimes slightly viscous with gland-tipped hairs. Panicle 6–12 × 2–4·5 cm., loose to somewhat contracted, ovate, elliptic, elliptic-oblong in outline, erect or rarely drooping; branches spreading to ascending, sometimes appressed to the rhachis; rhachis, branches and pedicels scabrous and often hairy. Spikelets up to 1·5 cm. long, pedicelled to subsessile, obovate to obovate-oblong in lateral view, sericeous to villous. Glumes light grey-green or dull green, acute to subobtuse, glabrous or finely pubescent, rarely with a few gland-tipped hairs, smooth or asperulous; the inferior (4·6)5–7·5(10·5) mm. long, (7)9–11(13)-nerved; the superior (6)6·5–9 (rarely –11·5) mm. long, (8)9–11 (rarely –14)-nerved, always slightly shorter than the spikelet. Lowest lemma (8·5)9·5–14(–18) mm. long; awns (4·5) 5·75–8(13) mm. long. Palea 4·25–5·75 mm. long, usually with a few gland-tipped hairs along the keels towards the apex or sometimes without glands. Anthers (2)2·5–3·5(4) mm. long. Caryopsis c. 1·5 mm. long, yellowish-brown.

Botswana. N: Francistown, 1040 m., 7.iii.1961, *Vesey-FitzGerald* 3114 (BM; SRGH). SW: c. 11·2 km. SW. of Kang on road to Murumush, 17.ii.1960, *de Winter* 7345 (K; SRGH). SE: Morale Pasture Station, Mahalapye, 980 m., iii.1957, *de Beer* 81854 (K; LISC; SRGH). **Zambia.** B: Senanga, Mashi R. bank near SW. corner, 1.xi.1964, *Verboom* 1506 (SRGH). S: Mazabuka, 6·4 km. from Chirundu Bridge on Lusaka road, 6.ii.1958, *Drummond* 5490 (K; LISC; SRGH). **Rhodesia.** N: Mtoko, N. Mtoko Reserve, 8.i.1958, *Cleghorn* 334 (BM; SRGH). W: Nyamandhlovu, Capt. Bury's farm, 24.ii.1932, *Rattray* 488 (BM; K; SRGH). C: Hartley, Poole Farm, 25.i.1944, *Hornby* 2353 (K; SRGH). E: Chipinga, 379 m., 19.i.1957, *Phipps* 14 (LISC; SRGH). S: Beitbridge, Chiturupazi, 25.ii.1961, *Wild* 5425 (BM; SRGH). **Malawi.** S: Port Herald, between Thangadzi and Lilanje R., c. 90 m., 25.iii.1960, *Phipps* 2705 (BM; K; SRGH). **Mozambique.** T: Tete, near the Campo de Aviação, c. 130 m., 12.iii.1964, *Torre & Paiva* 11180 (COI; LISC). MS: Chemba, Chiou, Experimental Station of C.I.C.A., 18.iv.1960, *Lemos & Macuácua* 119 (BM; COI; K; LISC; SRGH). SS: Caniçado, between Caniçado and Nalázi, 15.v.1948, *Torre* 7818 (LISC). LM: Magude, near Moine, 16.ii.1953, *Myre & Balsinhas* 1508 (K).

Also in Mauritania, Cape Verde Is., Socotra, West Africa, Sudan Republic, Angola, Uganda, Tanzania, Kenya, Somalia, Ethiopia, SW. Africa and S. Africa. In sandveld, woodland, open canopy woodland, on stony hillsides or sloping river banks, sometimes on lake-shore fringes, roadsides and around abandoned habitations, usually growing on sandy

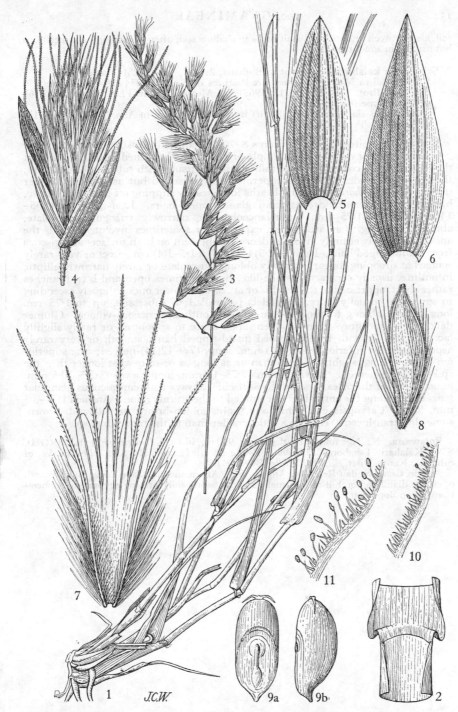

Tab. 40. SCHMIDTIA PAPPOPHOROIDES. 1, habit (×⅔); 2, ligule (×4); 3, inflorescence
(×⅔); 4, spikelet (×4); 5, inferior glume (×9); 6, superior glume (×9); 7, lemma
(×9); 8, palea (×9); 9, caryopsis (a) ventral (b) lateral views (×21); 10, 11, part
of the keel of the palea towards the apex (×24), all from *Lemos & Macuácua* 119.

soil, also between granite and basalt rocks in shallow soil, often forming pure associations but usually in small tussocks.

2. **Schmidtia kalahariensis** Stent in Bothalia, **2**: 421, 423 (1928) (" *kalihariensis* ").— Chippindall in Meredith, Grasses & Pastures of S. Afr.: 234 (1955).—Launert in Bol. Soc. Brot., Sér. 2, **39**: 315 (1965); in Merxm., Prodr. Fl. SW. Afr. **160**: 169 (1970). Type from S. Africa.

Antoschmidtia kalahariensis (Stent) Bremek. & Oberm. in Ann. Transv. Mus. **16**: 405 (1935). Type as above.

A coarse caespitose annual. Culms 8–70(90) cm. tall, prostrate, decumbent or geniculate, 3–6 (or more)-noded, simple or repeatedly branched, usually rooting at the lower nodes, coarsely striate, pubescent or hirsute with tubercle-based hairs; the lower internodes short, the uppermost long-exserted but usually less slender than in *S. pappophoroides*. Leaf-sheaths loose, easily slipping off the culm, striate, hirsute or pubescent, viscous from gland-tipped hairs. Leaf-laminae (1)2·5– 14(20) × (0·3)0·7–1·75 cm., linear-lanceolate or narrowly triangular-lanceolate, abruptly tapering to an acute apex, expanded or sometimes involute towards the apex, more rarely entirely involute, densely pubescent on both surfaces and viscous from gland-tipped hairs. Panicle (2·5)4–12(15) × 1·5–3(4) cm., erect or very rarely somewhat drooping, linear, narrowly oblong, lanceolate or rarely narrowly elliptic in outline, usually contracted, spike-like, but sometimes open and lax; branches rather short, appressed to the rhachis or at least the lower ones obliquely ascending to spreading, usually hairy. Spikelets pedicelled to subsessile, up to 1·75 cm. long, elliptic-oblong to obovate-triangular in outline, sericeous-villous. Glumes dark green, lead-grey-green or green, apex acute to subobtuse or rarely slightly lacerate, usually with tubercle-based gland-tipped hairs, smooth or very rarely asperulous; the inferior 4·8–9(–11) mm. long, (7)9–12(14)-nerved; the superior 6·6–15 mm. long, slightly shorter than or as long as or somewhat longer than the spikelet, 9–14-nerved. Lowest lemma 7·75–15 mm. long; awns (3·2)5–10·5 mm. long. Keels of the palea, besides the stiff cilia, always with conspicuous glandular hairs either along the entire keel or confined to the apical region. Anthers 1·75–3·5 mm. long. Caryopsis c. 2 mm. long, light golden-brown or yellowish-brown, somewhat translucent; embryo a little smaller than in the preceeding sp.

Botswana. N: Ngamiland, Matlapaneng, 940 m., iii.1958, *Robertson* 621 (K; SRGH). SW: Kalahari, Tshabong, 25.ii.1960, *Yalala* 86 (BM; SRGH). SE: 33·6 km. W. of Siherela, Kanye Distr., iii.1950, *Miller* B998 (K).

Also in Chad, Sudan Republic, Angola, SW. Africa and the Cape Prov. In grassland, open woodland and on hillsides; usually in poor sandy soil, forming almost pure associations, often dominant.

INDEX TO BOTANICAL NAMES

i